ADVENT[illegible] WITH MICROELECTRONICS

CW00495011

TOM DUNCAN

BPB PUBLICATIONS

B-14, CONNAUGHT PLACE, NEW DELHI-110001

HOW TO START

Microelectronics is about integrated circuits – or ICs as they are usually called. The ICs used here are much simpler than those in pocket calculators and digital watches but like them they consist of entire electronic circuits on a tiny 'chip' of silicon.

Before starting any of the projects you are recommended to read pages 4 to 9. This will acquaint you with the bits and pieces and with the 'breadboard' on which the projects are built. Your attention is drawn especially to the note about handling CMOS ICs on page 6.

The projects may be done in any order but it is suggested that you make your first one '1 Finding out about chips'. Practical layouts are always given but after you have assembled a few successfully you may find it more fun to work out your own arrangement from the circuit diagram. All circuits work from a 9 V battery.

Information about how to obtain a complete *Adventures with Microelectronics* kit or how to get the components separately is given on page 64. You will also need a pair of blunt-nosed pliers, a small screwdriver and either a pair of side-cutting pliers or wire-strippers.

WISHING YOU SUCCESSFUL ADVENTURES

CONTENTS

ACKNOWLEDGEMENTS

Brian scrutinized the circuits and made many helpful suggestions. Howard, Freda, John, Nicola, David and Fred also gave much appreciated assistance in various ways.

THE BITS AND PIECES

PART	CIRCUIT SIGN	WHAT IT DOES
Battery (9 V, type PP3)		Supplies a *voltage* which drives an electric *current* round the circuit from the positive (+) terminal of the battery to its negative (−) terminal. Voltage is measured in *volts* (V) and current in *amperes* (A).
Connecting wire (PVC-covered tinned copper wire 1/0.6 mm, i.e. 1 wire of diameter 0.6 mm)		Allows current to flow through it easily because it is made of copper which is a good electrical conductor. Insulators like PVC (polyvinyl chloride – a plastic) and enamel are used to cover connecting wires.
Miniature slide switch SPDT (single pole double throw)		Connects terminal A to terminal B or C, i.e. it is a change-over switch.
Resistor (carbon, ½ watt)		Reduces the current in a circuit because of its resistance. The coloured bands give the resistance in *ohms* (see page 9).
Photocell or **light dependent resistor** (l.d.r.) (e.g. ORP12)		When light falls on it, its resistance becomes small; in the dark its resistance is high.

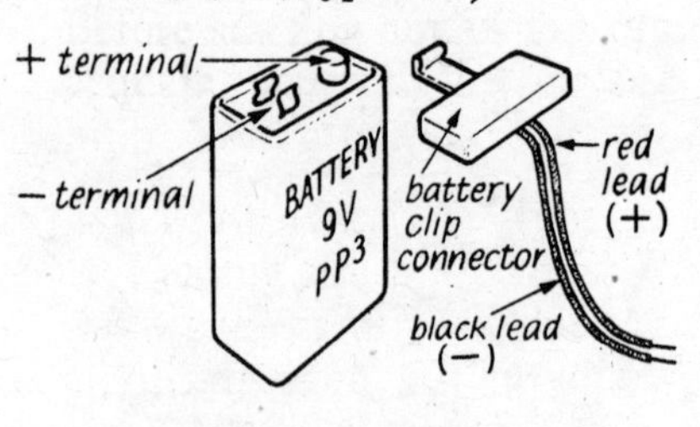

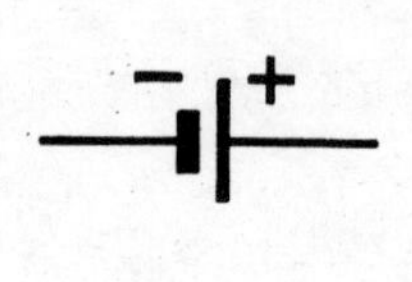

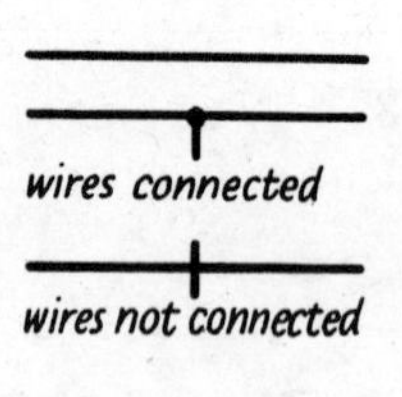

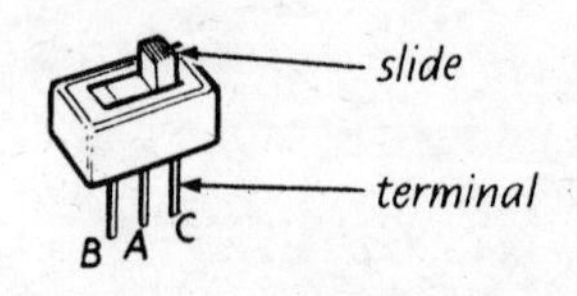

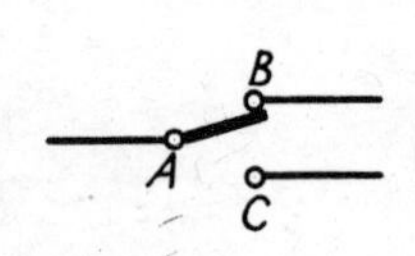

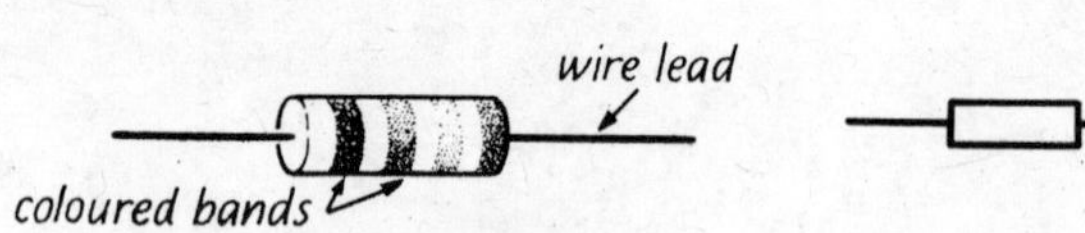

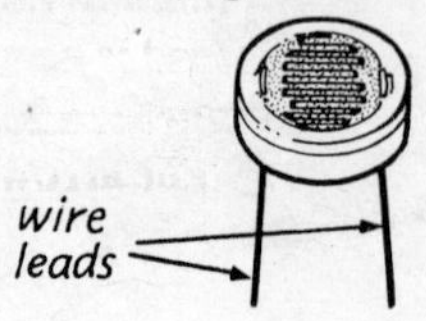

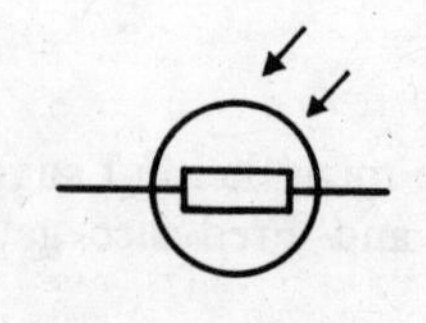

PART | 

| **WHAT IT DOES**

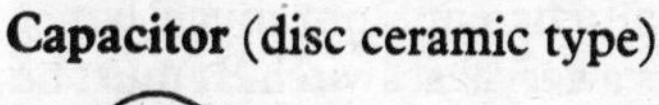

Capacitor (disc ceramic type)

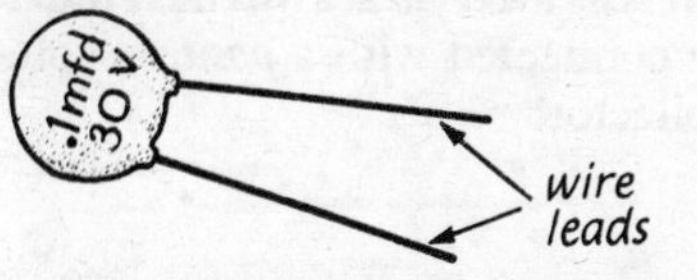

Stores electricity; the greater the capacitance the more does it store. Capacitance values are measured in microfarads shortened to μF or, less correctly, to mfd. On a capacitor, 0.1 μF may be marked as .1 mfd and 0.01 μF as 10 n. The greatest voltage it can stand is also shown, e.g. 30 V.

Electrolytic capacitor

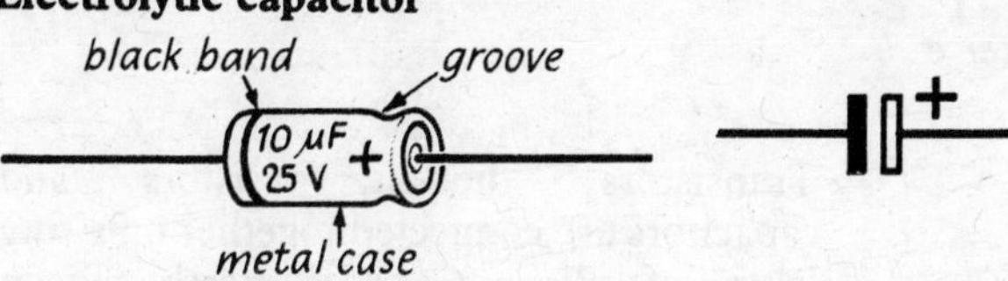

Stores electricity: values usually larger than 1 μF. Greatest voltage marked on it. *Must be connected the correct way round.*

Variable capacitor
(0.0005 microfarads)

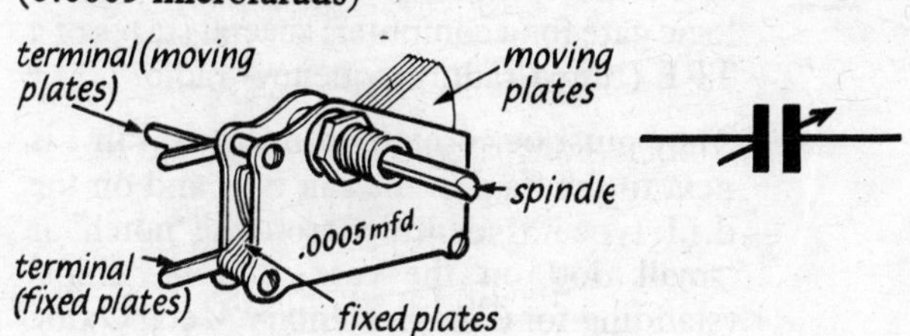

Varies the capacitance in a circuit by moving one set of metal plates in or out of a fixed set when the spindle is rotated. The sets of plates are separated by sheets of an insulator (also called a dielectric).

Loudspeaker (2½ inch,
25 to 80 ohms)

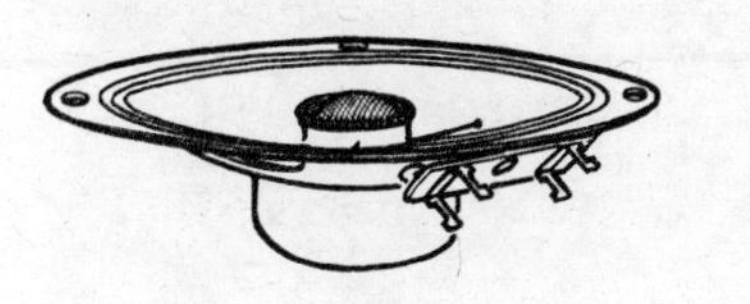

Changes electric currents into sound.

Ferrite rod aerial

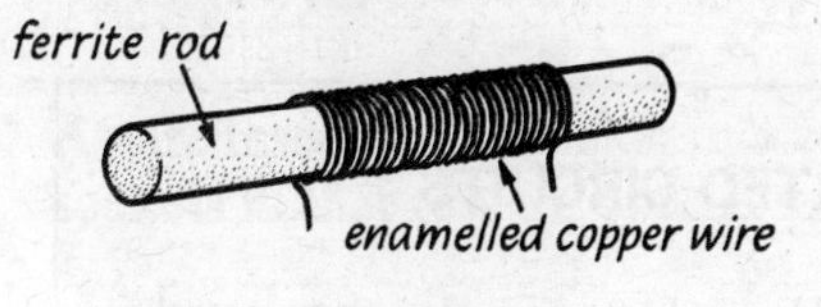

Changes radio waves into electric currents.

LED (light emitting diode)

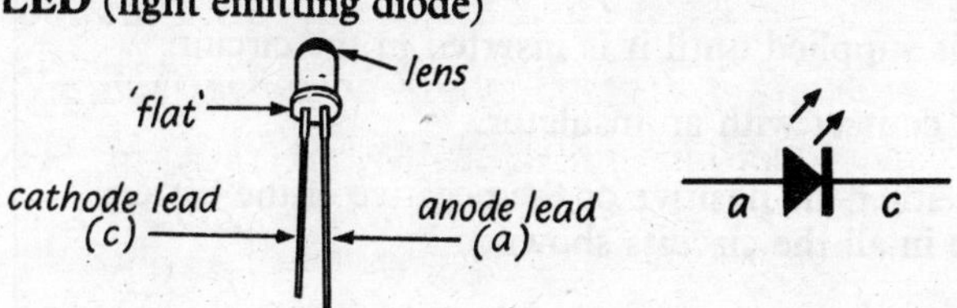

Lets current flow in one direction but not in the other. When it conducts, light is emitted. *Must have a current limiting resistor in series with it.* The cathode lead is nearest the 'flat' and *may* be shorter than the anode lead (but this is not always so). The arrow on the sign shows the conducting direction.

PART	CIRCUIT SIGN	WHAT IT DOES
Transistor (npn) (e.g. ZTX300 or 2N3705)		Amplifies small currents into much larger copies. Acts as a very fast switch. It must be correctly connected with a *positive* voltage to the collector.

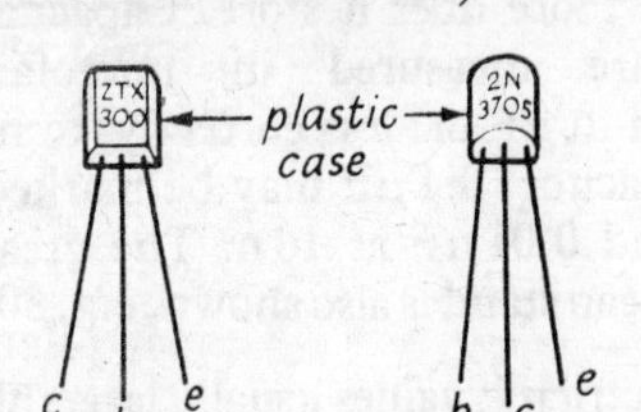

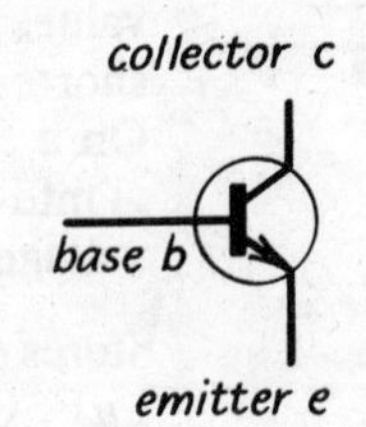

Integrated circuits ('chips')

1 *Can and plastic types*

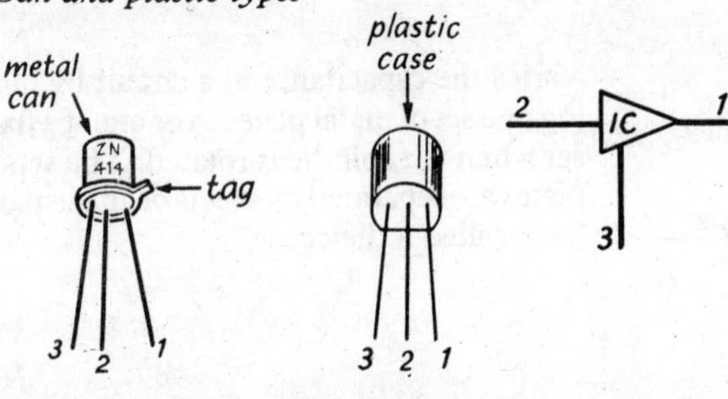

2 *Dual-in-line* (d.i.l) type (14 or 16 pins)

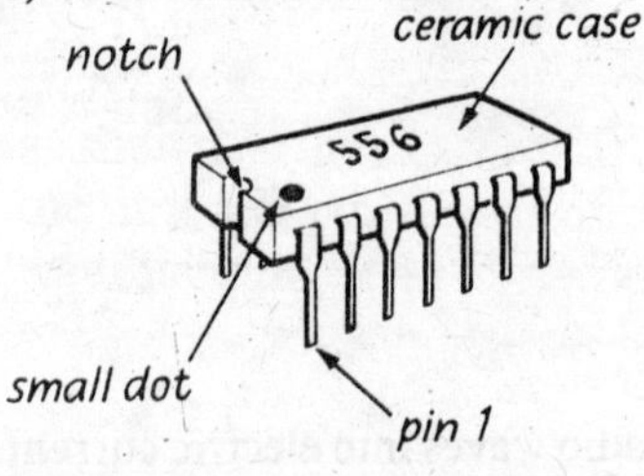

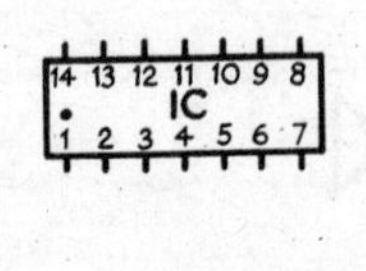

Transistors, diodes, resistors and capacitors are connected together on a tiny 'chip' of silicon (sand is mostly silicon oxide) to give any desired circuit, e.g. a multistage amplifier; an astable, bistable or monostable multivibrator; a counter; a logic gate for a computer; several stages of a TRF (tuned radio frequency) radio.

They must be correctly connected. Pin 1 is next to the 'tag' in the can type and on the d.i.l. type it is identified from the 'notch' or 'small dot' on the case. CMOS 'chips' (standing for *C*omplementary *M*etal *O*xide *S*emiconductors and pronounced 'see-moss') need special care.

PRECAUTIONS WITH CMOS INTEGRATED CIRCUITS

Damage occurs if static charges build up on input pins when, for example, they touch insulating materials (e.g. clothes, plastic pen) in warm, dry conditions.

1 Keep the IC in the carrier in which it is supplied until it is inserted in the circuit.

2 Do not finger the pins or hold them in contact with an insulator.

3 Connect all unused inputs of the IC to either the positive or the negative of the battery, depending on the circuit. (This is done in all the circuits shown.)

BUILDING CIRCUITS

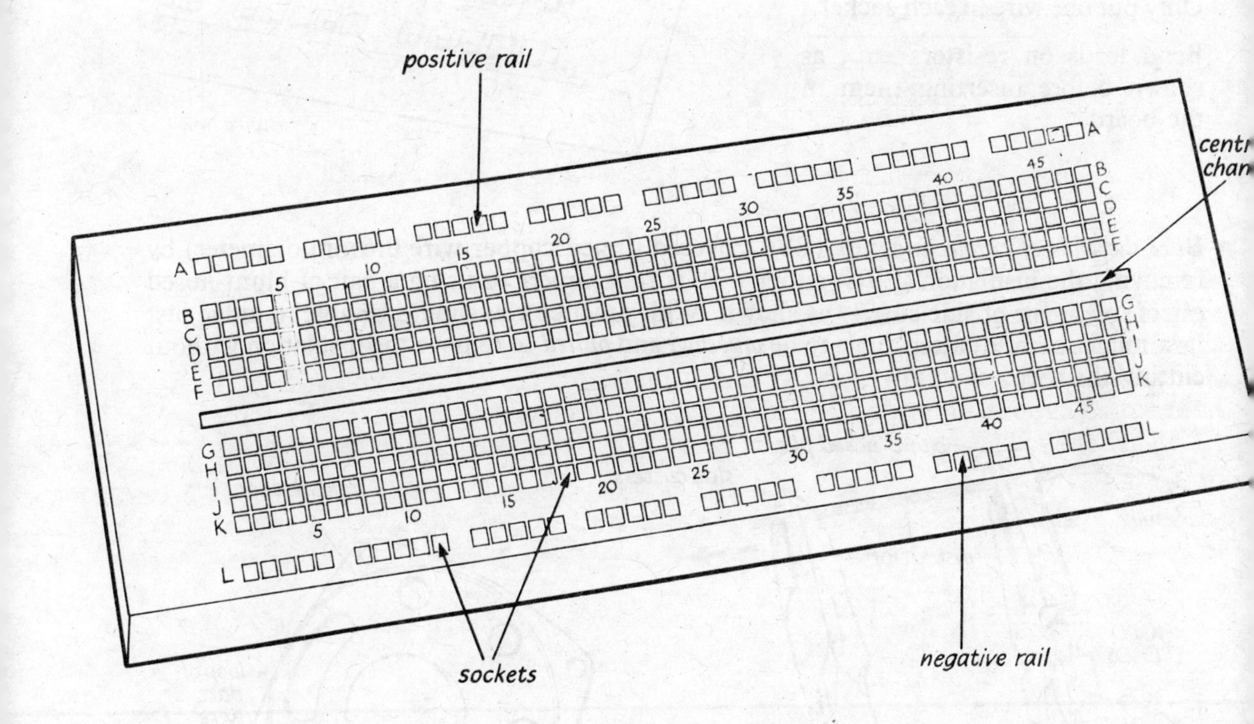

The circuit board shown above accepts ICs as well as separate components. It has 47 rows of 5 interconnected sockets on each side of a central channel across which d.i.l. ICs can be fitted. A wire inserted in a socket in a certain row becomes connected to wires in any of the other 4 sockets in that row by a metal strip under the board. For example, wires in sockets B5, C5, D5, E5 and F5 (shown in colour in the diagram) are all joined.

A row of 40 interconnected sockets along the top of the board and a similar row along the bottom act as the positive and negative power supply rails (called 'bus bars').

Various makes of circuit board are available, some with vertically mounting removable panels for supporting controls (see page 63).

1 *To make a connection* push about 1 cm of the *bare* end of a wire (0.25 to 0.85 mm diameter) *straight* into the socket (not at an angle) so that it is gripped by the metal strip under the board. Do not use wires that are *dirty* or have *kinked* ends. Only put one wire in each socket.

Bend leads on resistors etc., as shown before inserting them in the board.

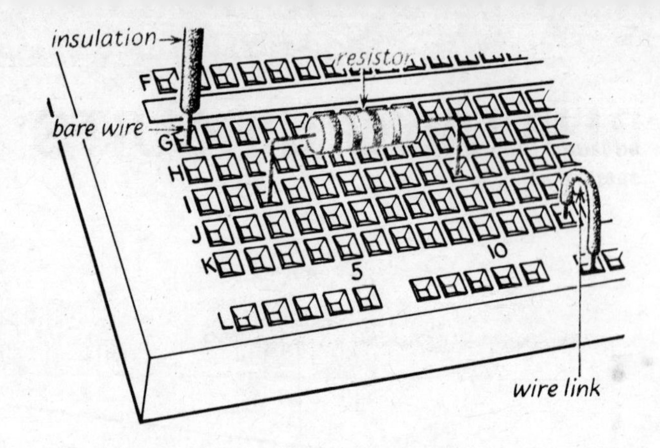

2 *Bare the ends of connecting wire* (PVC-covered tinned copper wire 0.6 mm diameter) by removing the insulation (PVC) either with wire strippers or using a pair of blunt-nosed pliers and a pair of side cutters as shown. With practice you should be able to judge just how much the side cutters have to be *squeezed* and *pulled* to remove the insulation without cutting the wire.

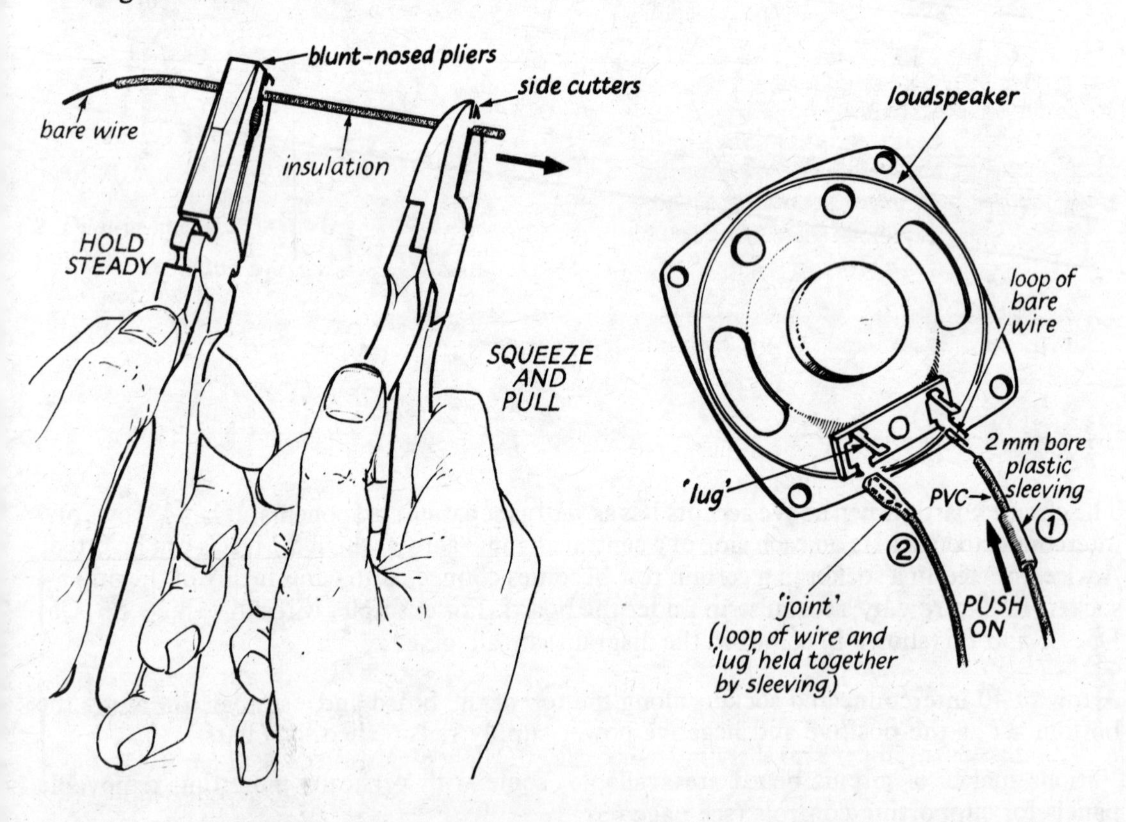

3 *'Join' wires to the 'lugs'* on the loudspeaker and variable capacitor using a small length of 2 mm bore plastic sleeving – as shown by 1 and 2 .

RESISTOR COLOUR CODE

Resistor values are given in ohms (shortened to Ω, the Greek letter 'omega'). They are marked on the resistor using a colour code.

Three coloured bands are painted round the resistor. Each colour stands for a number. To read the colour code, start at the 1st band; it is nearest the end. Sometimes it is not clear which is the 1st band because there is a 4th band of gold or silver near the other end. These two colours are not used for the 1st band; they give the accuracy of the resistor (gold is ± 5% and silver ± 10%), so you should not have too much trouble deciding where to start.

The *1st band* gives the first number, the *2nd band* gives the second number and the *3rd band* tells how many noughts come after the first two numbers.

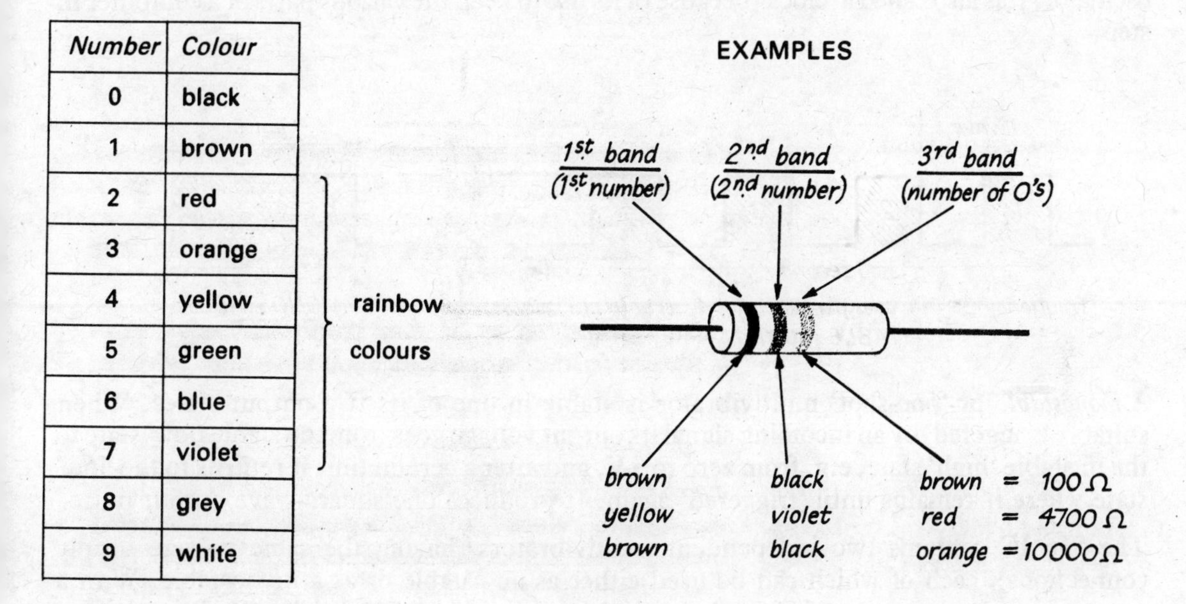

Number	Colour
0	black
1	brown
2	red
3	orange
4	yellow
5	green
6	blue
7	violet
8	grey
9	white

More resistor shorthand

1000 Ω = 1 kilohm = 1 kΩ
4700 Ω = 4.7 kilohm = 4.7 kΩ
1 000 000 Ω = 1 megohm = 1 MΩ

1 FINDING OUT ABOUT 'CHIPS'

By working through this introductory 'project' you will get a basic knowledge of what different integrated circuits ('chips') do. You will then understand better the projects that follow.

(A) Dual astable/monostable multivibrator (556)

A multivibrator has two output states – 'high' (when the output voltage equals that of the battery, e.g. 9 V) and 'low' (when the output voltage is zero). An *astable* or 'free-running' multivibrator is stable in neither state (hence 'astable' which means 'not stable') but switches to and fro from one state to the other to give a square-wave output, i.e. it is a square-wave oscillator. It is also called a 'clock' because of its use to keep the various parts of a computer in step.

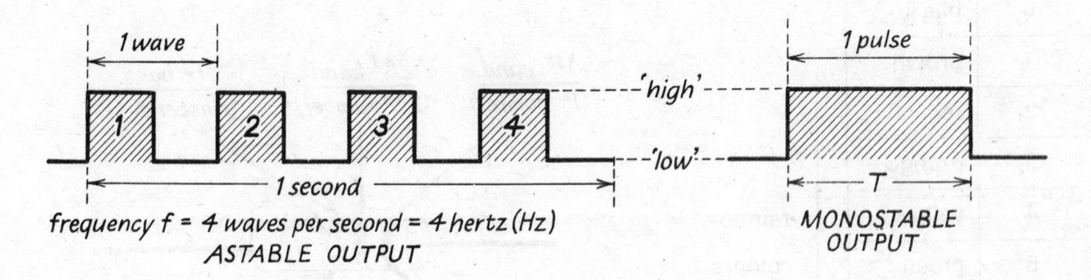

A *monostable* or 'one-shot' multivibrator is stable in one of its two output states. When suitably 'triggered' by an incoming signal its output voltage goes from the stable 'low' state to the unstable 'high' state, e.g. from zero to 9 V, and after a certain time it returns to the 'low' state where it remains until 'triggered' again. It produces one square-wave output pulse.

The 556 IC contains two independent multivibrators (having the same voltage supply connections), each of which can be used either as an astable or as a monostable. With a maximum output current of 200 mA it can drive a loudspeaker as well as LEDs.

WHAT YOU NEED

Dual astable/monostable multivibrator IC (556); red LED (see *Parts list*, p. 63); resistors – 680 Ω (blue grey brown), three 10 kΩ (brown black orange), 33 kΩ (orange orange orange), 220 kΩ (red red yellow), 1 MΩ (brown black green), 2.2 MΩ (red red green); electrolytic capacitors – two 1 μF, 4.7 μF; disc ceramic capacitors – 0.01 μF, 0.1 μF; loudspeaker 2½ in, 25 to 80 Ω; 9 V battery; battery clip connector; circuit board; PVC-covered tinned copper wire 1/0.6 mm.

HERE IS THE ASTABLE CIRCUIT

Two external resistors $R1$ and $R2$ and one external capacitor $C1$ are required. The frequency f is given by

$$f = \frac{1.4}{(R1 + 2 \times R2)\ C1} \text{ Hz}$$

where $R1$ and $R2$ are in ohms and $C1$ in farads. If $R2$ is much greater than $R1$ (as here) then

$$f = \frac{1.4}{2 \times R2 \times C1} = \frac{0.7}{R2 \times C1}$$

For example, if $R2 = 1\ \text{M}\Omega = 10^6\Omega$ and $C1 = 1\ \mu\text{F} = 10^{-6}\text{F}$ then $f = 0.7/(10^6 \times 10^{-6}) = 0.7$ Hz which is about 1 square wave per second.

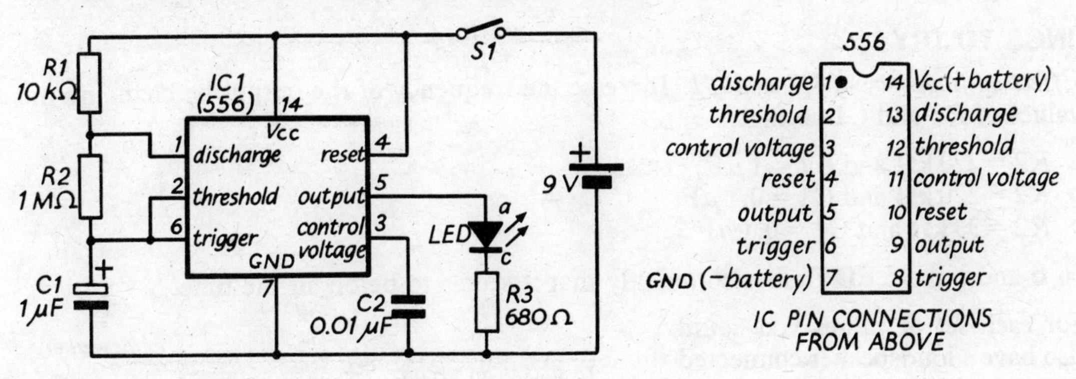

HOW TO BUILD IT

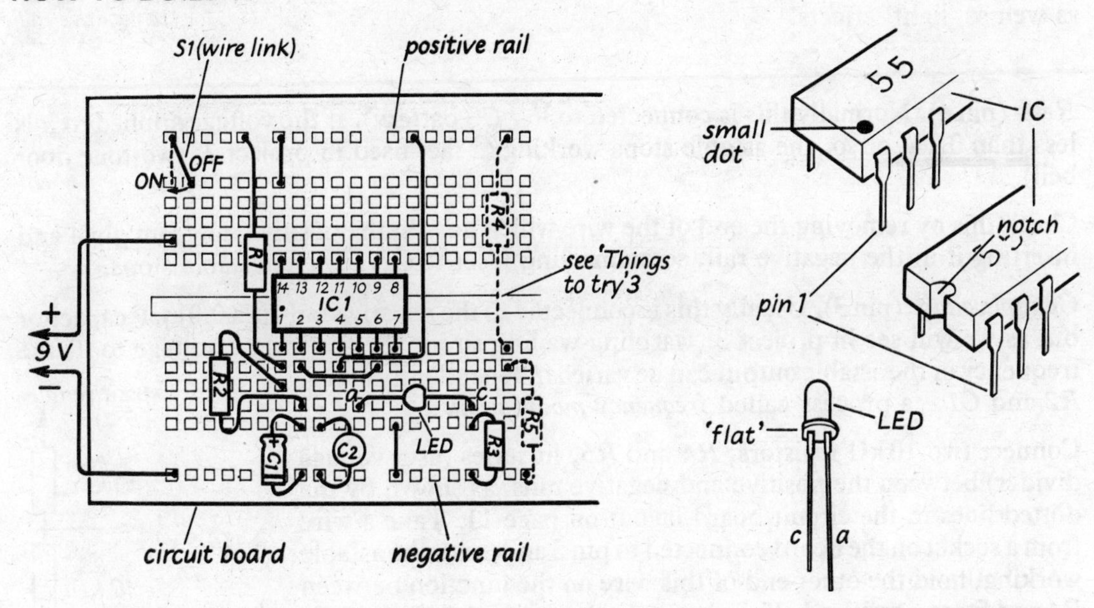

1 Identify pin 1 on the IC from the small dot or notch at one end of the case. Carefully push the IC into the circuit board in the position shown, taking care not to bend any of the pins.

2 Insert wire links from the IC to the positive and negative rails and between other sockets on the board, as shown.

3 Insert $R1$, $R2$, $R3$, $C1$ and $C2$. Be sure the electrolytic capacitor $C1$ is the correct way round, the + end has a groove and the − end a black band round it (usually).

4 Insert the LED remembering that the cathode is nearest the 'flat' at the base of the case (or the cathode lead *c* may be shorter than the anode lead *a*).

5 CHECK THE CIRCUIT CAREFULLY.

6 Connect the battery with the correct polarity. The wire link $S1$ acts as the battery on/off switch. Switch $S1$ on. The LED should flash about once a second (showing that the astable is working); if it doesn't it may be connected the wrong way round.

THINGS TO TRY

1 *Effect on frequency of R2 and C1.* Increase the frequency of the output by changing the values of $R2$ and $C1$ so that

a $R2 = 220\,k\Omega$ and $C1 = 1\,\mu F$,
b $R2 = 220\,k\Omega$ and $C1 = 0.1\,\mu F$,
c $R2 = 33\,k\Omega$ and $C1 = 0.1\,\mu F$.

In **b** and **c** the LED flashes so quickly that it seems to be on all the time.

For each set of values you could also have a loudspeaker connected to the astable output (pin 5) via a 1 μF capacitor so getting 'sound' as well as 'light' effects.

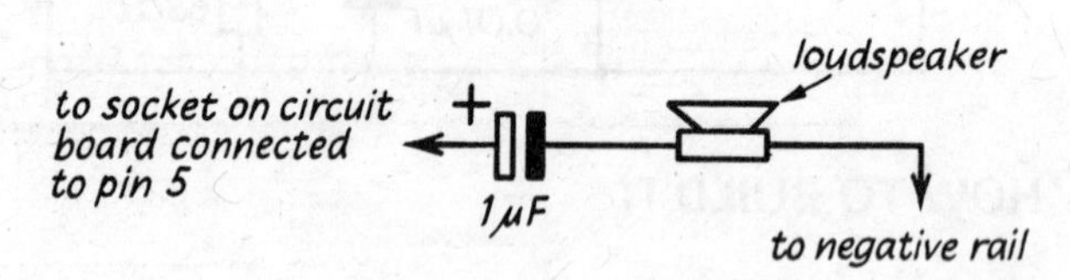

2 *Reset* (pin 4). Normally this is connected to V_{CC} (+ battery). If the voltage applied to it is less than 0.7 V or so, the astable stops working (a fact used in project 2: two-tone door bell).

Check this by removing the end of the wire which goes to the positive rail from pin 4 and inserting it in the negative rail, so connecting reset to *GND*. The astable stops.

3 *Control voltage* (pin 3). Usually this is connected to the negative rail via a 0.01 μF capacitor but as we will see in project 3 (warbling-wailing siren), by applying a voltage to it, the frequency of the astable output can be varied independently of $R1$, $R2$ and $C1$ – a process called *frequency modulation.*

Connect two 10 kΩ resistors, $R4$ and $R5$, in series (as a voltage divider) between the positive and negative rails, as shown by the dotted lines in the circuit board layout on page 11. Take a wire from a socket on the board connected to pin 3 and, with the astable working, hold the other end of this wire on the junction between $R4$ and $R5$ so applying half the battery voltage (i.e. 4.5 V) to pin 3. The astable frequency increases – as shown by increases in the LED flashing rate and the pitch of the note from the loudspeaker.

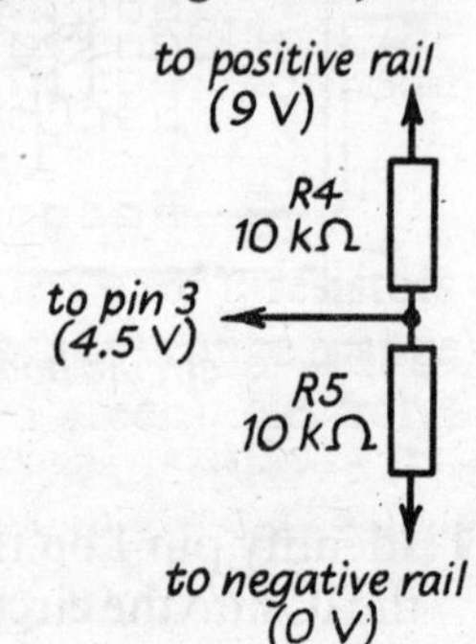

HERE IS THE MONOSTABLE CIRCUIT

One external resistor $R1$ and one external capacitor $C1$ are needed. The time T for which the pulse lasts is given by

$$T = 1.1 \times R1 \times C1$$

where T is in seconds if $R1$ is in ohms and $C1$ in farads. For example if $R1 = 2.2\,\text{M}\Omega = 2.2 \times 10^6\,\Omega$ and $C1 = 1\,\mu\text{F} = 10^{-6}\,\text{F}$ then $T = 1.1 \times 2.2 \times 10^6 \times 10^{-6} = 2.4\,\text{s}$.

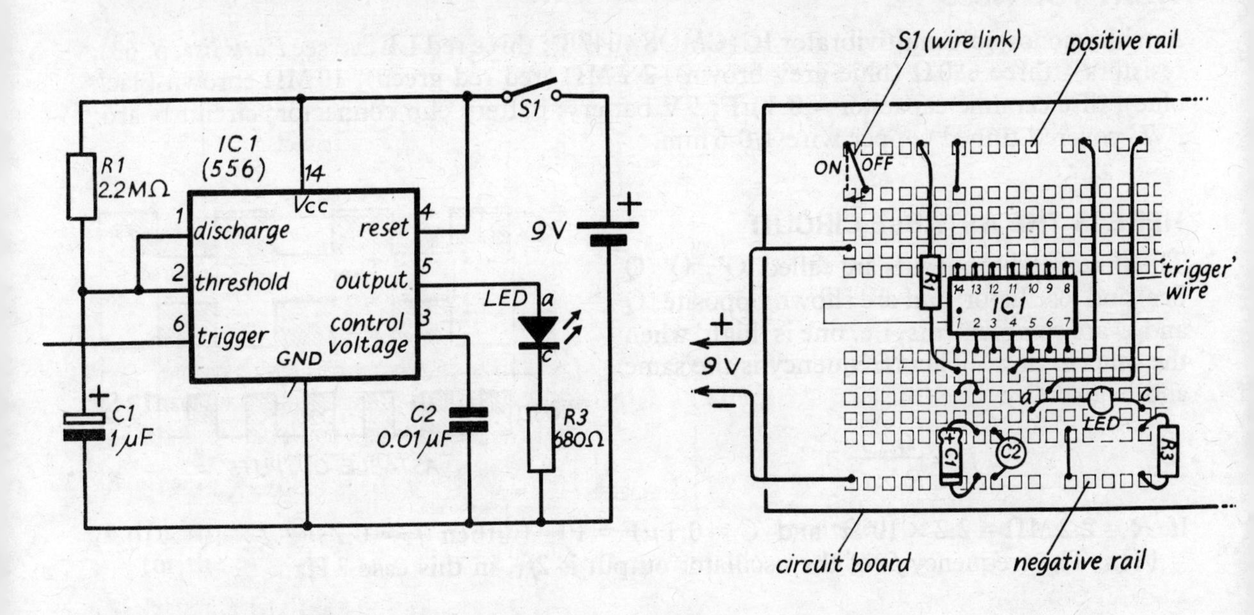

HOW TO BUILD IT

Make the following changes to the astable circuit.

1. Remove $R2$.
2. Change $R1$ from 10 kΩ to 2.2 MΩ.
3. Remove the wire link from pins 2 and 6 and use it to join pins 1 and 2.
4. Take a 10 cm long wire from pin 6 (trigger) to the positive rail.

The monostable is triggered by the *negative* (falling) edge AB of a pulse. To do this, switch $S1$ on, remove the 'trigger' wire from the positive rail and push the end momentarily into the negative rail, then return it *at once* to the positive rail. The LED should light up for about 2 seconds. So long as the triggering time t is less than the monostable output pulse time T (see page 10) the monostable works.

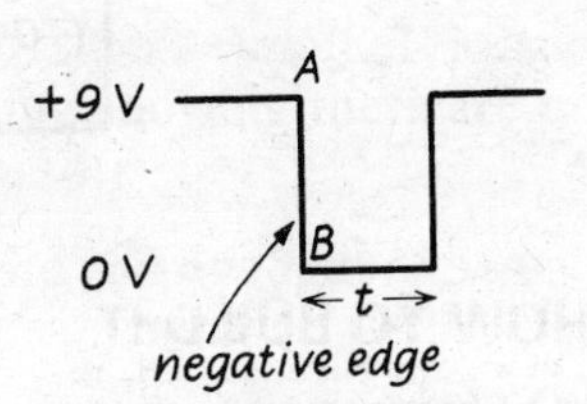

THINGS TO TRY

Change $C1$ from 1 μF to 4.7 μF – the output pulse should last for about 10 seconds.

(B) CMOS astable/monostable multivibrator (4047B)

This is a lower power multivibrator than the 556 (page 10) but it needs only one external resistor and capacitor (which must *not* be an electrolytic) and has three astable outputs. Each output can drive an LED but to operate a loudspeaker a transistor amplifier is used. In monostable operation it can be triggered by positive or negative edges.

WHAT YOU NEED

Astable/monostable multivibrator IC (CMOS 4047B); three red LEDs (see *Parts list*, p. 63); resistors – three 680 Ω (blue grey brown), 2.2 MΩ (red red green), 10 MΩ (brown black blue); disc ceramic capacitor – 0.1 μF; 9 V battery; battery clip connector; circuit board; PVC-covered tinned copper wire 1/0.6 mm.

HERE IS THE ASTABLE CIRCUIT

The three astable outputs are called 'Q', '$\bar{Q}$' (Q bar) and 'oscillator' and are shown opposite. Q and $\bar{Q}$ are complements, i.e. one is 'high' when the other is 'low'. Their frequency is the same and is given by

$$f_1 = \frac{0.23}{R \times C}$$

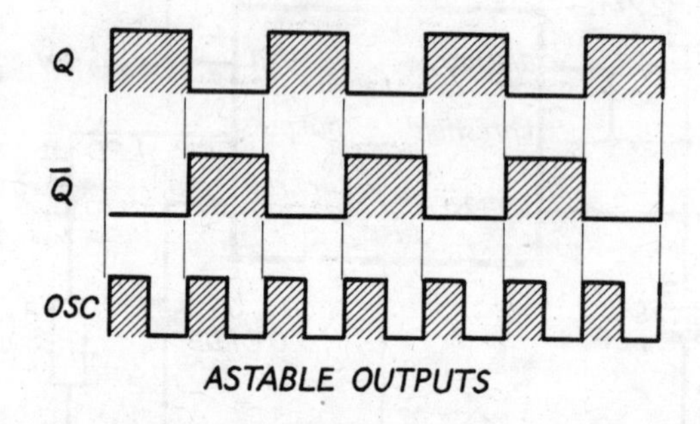

ASTABLE OUTPUTS

If $R = 2.2\,\text{M}\Omega = 2.2 \times 10^6\,\Omega$ and $C = 0.1\,\mu\text{F} = 10^{-7}\,\text{F}$ then $f_1 = 0.23/(2.2 \times 10^6 \times 10^{-7}) = 1\,\text{Hz}$. The frequency f_2 of the oscillator output is $2f_1$, in this case 2 Hz.

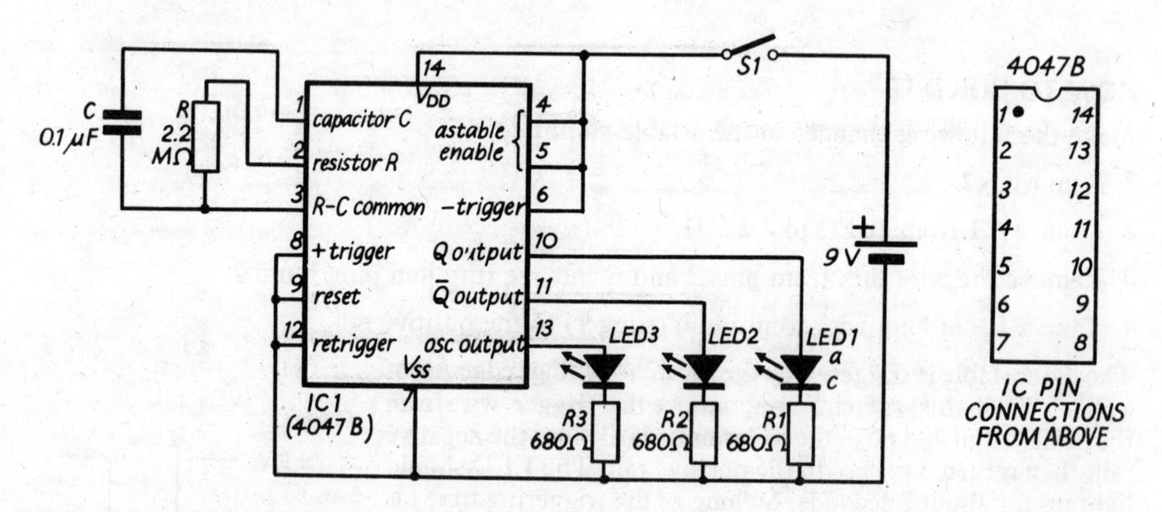

HOW TO BUILD IT

1 Identify pin 1 on the IC from the small dot or notch at one end of the case. Carefully push the IC into the circuit board in the position shown, taking care not to bend or touch any of the pins.

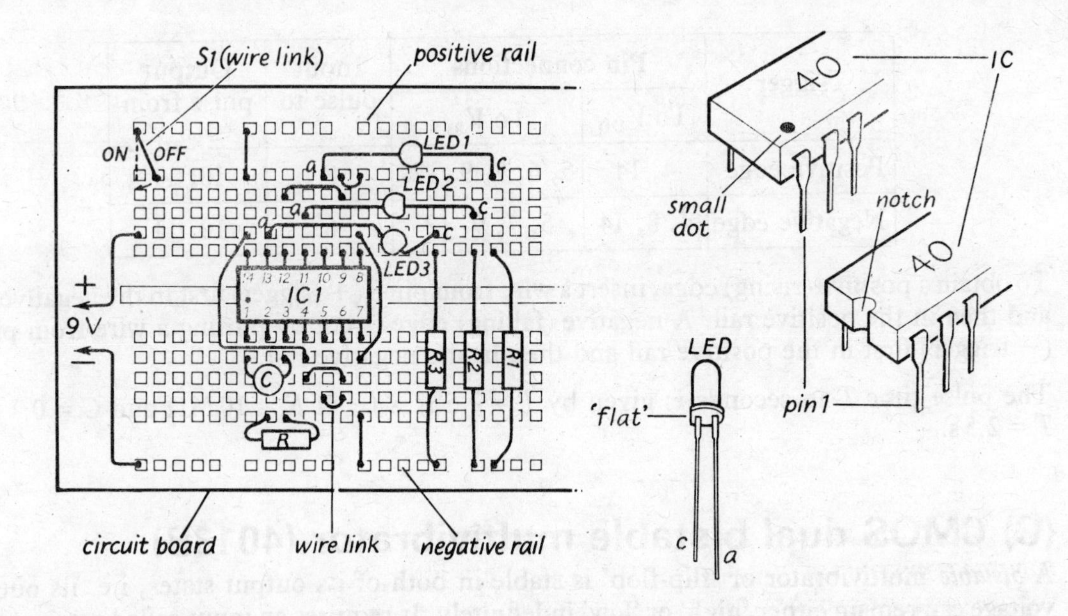

2 Insert wire links from the IC to the positive and negative rails and between other sockets on the board, as shown.

3 Insert *R*, *R*1, *R*2, *R*3 and *C*.

4 Insert the LEDs remembering that the cathode is nearest the 'flat' at the base of the case (or the cathode lead *c* may be shorter than the anode lead *a*).

5 CHECK THE CIRCUIT CAREFULLY.

6 Connect the battery with the correct polarity. The wire link *S*1 acts as the battery on/off switch. Switch *S*1 on. The LEDs should start flashing on and off; if any don't it may be because they are the wrong way round. Notice that (*a*) when LED 1 (Q output) is on, LED 2 ($\overline{Q}$ output) is off and vice versa and (*b*) LED 3 (osc output) flashes twice as fast as LEDs 1 and 2. If the flashing is too fast for you to follow, change *R* from 2.2 MΩ to 10 MΩ.

THINGS TO TRY

1 *Astable enable* (pin 5). Normally this is connected to V_{DD} (+ battery). If the voltage to it is less than a certain value (see project 6: light-operated alarm, page 38, *Things to try 1*) the astable does not work.

To check this, remove the small wire link connecting pins 5 and 6 and take another, longer wire from pin 5 to the negative rail. The LEDs will stop flashing showing there are no longer square-wave outputs from the astable.

Now remove the long wire from the negative rail and put it in the positive rail – this will *enable* the astable to work again.

2 *Monostable operation*. We do not use this IC as a monostable in any of our projects but if you want to try it in this method of operation the connections for positive and negative edge triggering are shown in the table overleaf.

Trigger	Pin connections		Input pulse to	Output pulse from
	To V_{DD}	To V_{SS}		
Positive edge	4, 14	5, 6, 7, 9, 12,	8	10, 11
Negative edge	4, 8, 14	5, 7, 9, 12	6	10, 11

To obtain a positive (rising) edge, insert a wire from pin 8 (+ trigger) first in the negative rail and then in the positive rail. A negative (falling) edge is got by putting a wire from pin 6 (− trigger) first in the positive rail and then in the negative rail.

The pulse time T in seconds is given by $T = 2.5\ R \times C$. If $R = 10\ \mathrm{M\Omega}$ and $C = 0.1\ \mu\mathrm{F}$, $T = 2.5\ \mathrm{s}$.

(C) CMOS dual bistable multivibrator (4013B)

A *bistable* multivibrator or 'flip-flop' is stable in both of its output states, i.e. its output voltage can remain either 'high' or 'low' indefinitely. It requires an input called a 'trigger' to change the state. Bistables can be connected to form an electronic counter which counts the number of input 'triggers' – as explained later (see page 17, *How to build it 6*).

The 4013B is a dual bistable with common voltage pins.

WHAT YOU NEED

Dual bistable multivibrator IC (CMOS 4013B); two red LEDs (see *Parts list*, p. 63); resistors – 680 Ω (blue grey brown), 4.7 kΩ (yellow violet red); miniature slide switch SPDT; 9 V battery; battery clip connector; circuit board; PVC-covered tinned copper wire 1/0.6 mm.

HERE IS THE CIRCUIT

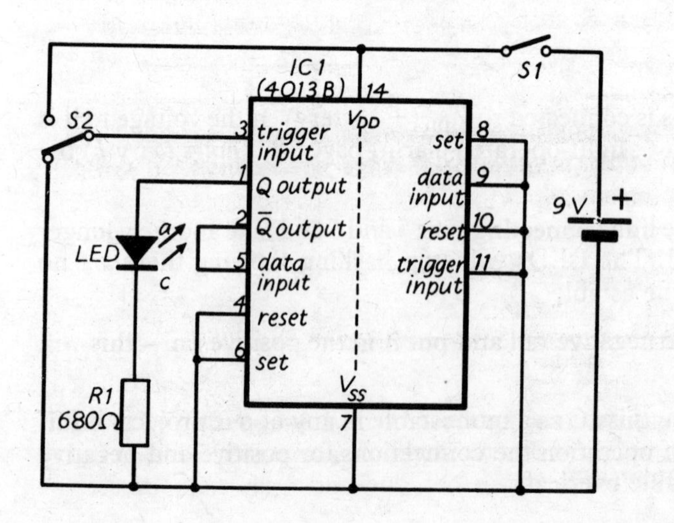

FIRST BISTABLE	4013 B		SECOND BISTABLE
Q output	1•	14	V_{DD} (+battery)
$\bar{Q}$ output	2	13	Q output
trigger input	3	12	$\bar{Q}$ output
reset	4	11	trigger input
data input	5	10	reset
set	6	9	data input
V_{SS} (−battery)	7	8	set

IC PIN CONNECTIONS FROM ABOVE

HOW TO BUILD IT

1 Identify pin 1 on the IC from the small dot or notch at one end of the case. Carefully push the IC into the circuit board in the position shown, taking care not to touch or bend the pins.

2 Insert wire links from the IC to the positive and negative rails and between other sockets on the board, as shown.

3 Insert *R*1 and *S*2. Set the switch to the *right* on *S*2.

4 Insert the LED remembering that the cathode is nearest the 'flat' at the base of the case (or the cathode lead *c* may be shorter than the anode lead *a*).

5 CHECK THE CIRCUIT CAREFULLY.

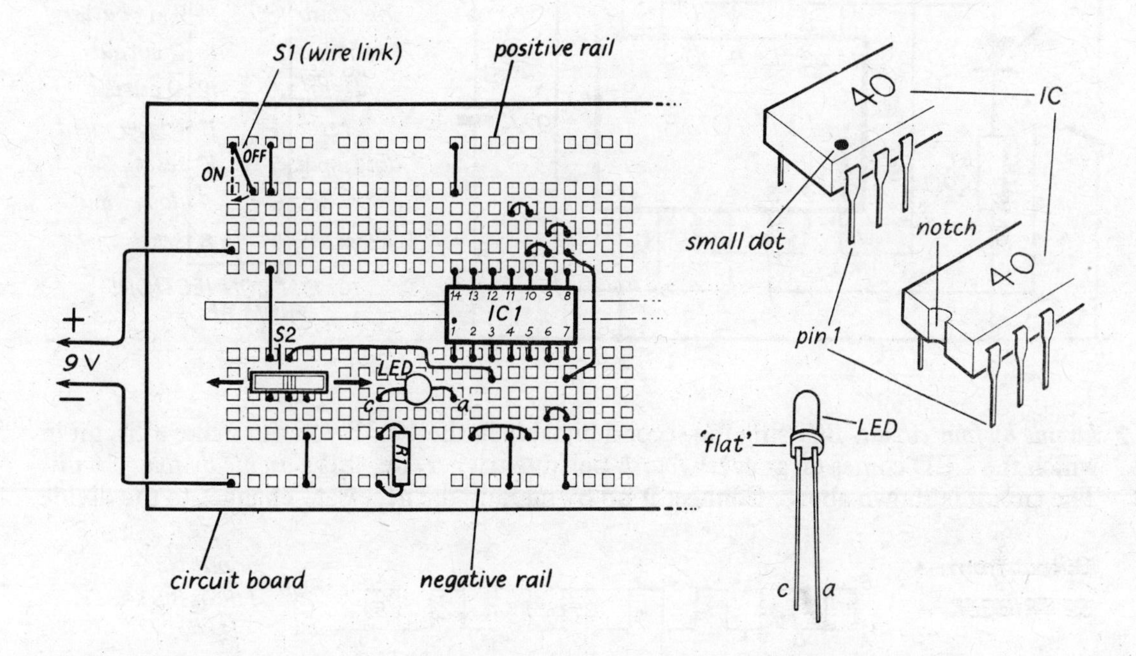

6 Connect the battery with the correct polarity. The wire link acts as the battery on/off switch *S*1. Switch *S*1 on. If the LED does not light up, switching *S*2 to the left and then back to the right should bring it on. (If it doesn't the LED may be connected the wrong way round.)

Every time *S*2 is switched from right to left, the voltage at trigger input (pin 3) rises from zero to 9 V and this *positive* (rising) edge makes the bistable change its state, i.e. if the output is 'high' (LED alight), it goes 'low' (LED off) and vice versa.

Starting with the LED off, apply *four* positive triggers to the bistable, i.e. switch *S*2 to the left and then to the right *four* times. You should find that the LED comes on only twice, i.e. the Q output of the bistable is 'high' just *twice*. The bistable counts every *second* positive trigger input; it is a *divide by two* circuit. Check that this is so for six, eight, etc. positive triggers.

THINGS TO TRY

1 $\bar{Q}$ *output*. Connect another LED and a 4.7 kΩ current-limiting series resistor between the $\bar{Q}$ (Q bar) output (pin 2) and the negative rail. When you apply positive triggers now you will find that

a the Q output is 'high' when the $\bar{Q}$ output is 'low' and vice versa, i.e. the Q and $\bar{Q}$ outputs are complements.

b the frequency of the $\bar{Q}$ output pulses is (like the Q output) half the frequency of the positive triggers.

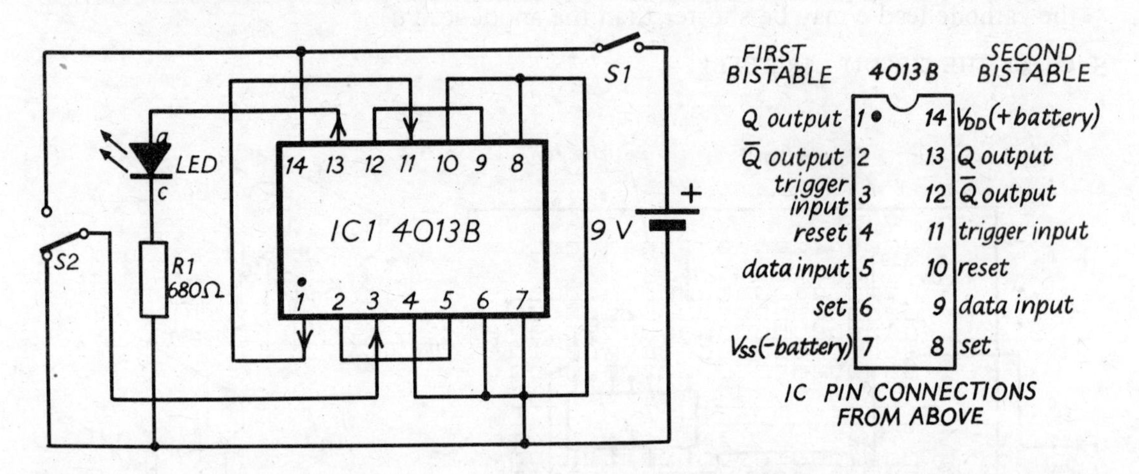

2 *Divide by four circuit.* By using the second bistable in the 4013B you can make a circuit in which the LED comes on at every *fourth* positive trigger, i.e. it is a *divide by four* circuit. The circuit is shown above. Connect it up by making the necessary changes to the divide

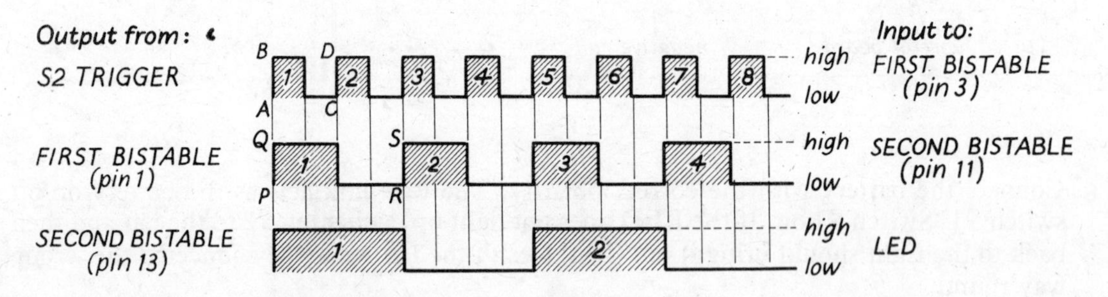

by two circuit. In the above diagram the first line shows the 'triggers' obtained by switching *S*2 between battery − and + (i.e. zero and 9 V). These are fed to the trigger input of the first bistable (pin 3) and change its output state (from 'low' to 'high' or vice versa) on *positive* edges such as AB, CD, etc. The second line shows the resulting Q output (at pin 1) of the first bistable. This output (which is fed to the trigger input of the second bistable, i.e. pin 11) causes the second bistable to change its output state (from 'low' to 'high' or vice versa) on *positive* edges such as PQ, RS, etc. The third line shows the output from the second bistable (pin 13) which is fed to the LED and lights it up on 'high'.

(D) CMOS quad two-input NAND gate (4011B)

Computers contain switches called *logic gates* which 'open' and give a 'high' output voltage only when certain conditions are satisfied at their inputs (of which they have more than one). There are four main types of gate – AND, NAND, NOR and OR. In their 'truth tables' given below for two inputs A and B, a 'high' output or input (e.g. 9 V) is shown by a '1' and a 'low' output or input (e.g. 0 V) by a '0'.

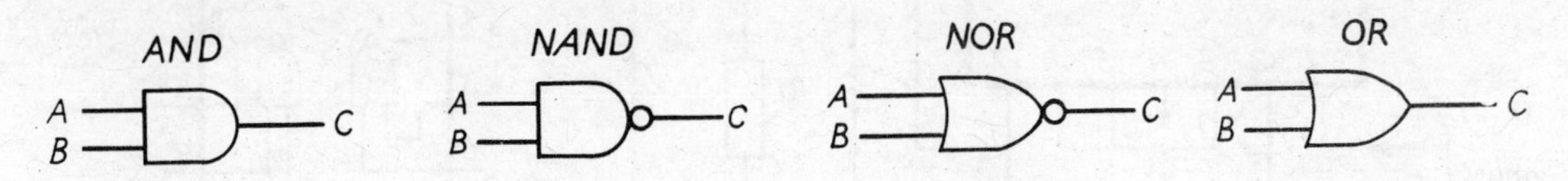

AND

Inputs		Output
A	B	C
0	0	0
0	1	0
1	0	0
1	1	1

NAND

Inputs		Output
A	B	C
0	0	1
0	1	1
1	0	1
1	1	0

NOR

Inputs		Output
A	B	C
0	0	1
0	1	0
1	0	0
1	1	0

OR

Inputs		Output
A	B	C
0	0	0
0	1	1
1	0	1
1	1	1

An AND gate gives a 'high' output only when *both* its inputs are 'high'; a NOR gate gives a 'high' output only when *both* its inputs are 'low'. We will use both of these gates in later projects.

Each type of gate can be made from a combination of one type of the others. For example, AND, OR and NOR gates can all be built from NAND gates, as shown below.

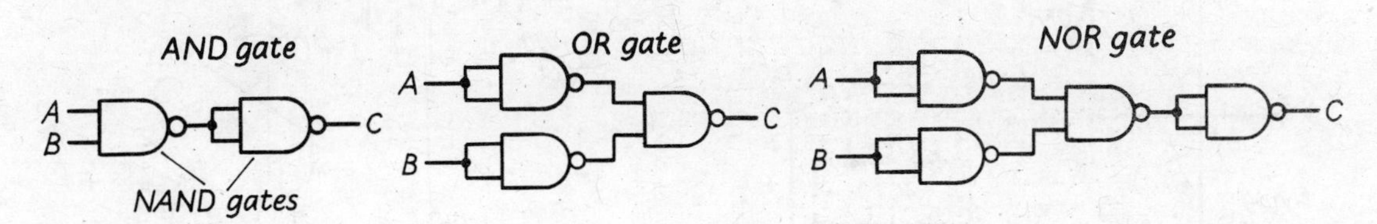

The 4011B is a quad 2-input NAND gate, i.e. it has four NAND gates each with two inputs and having the same voltage supply pins. We will use it to make the other three gates.

WHAT YOU NEED

Quad 2-input NAND gate IC (CMOS 4011B); red LED (see *Parts list*, p. 63); resistor – 680 Ω (blue grey brown); 9 V battery; battery clip connector; circuit board; PVC-covered tinned copper wire 1/0.6 mm.

HERE IS THE 'AND' GATE CIRCUIT

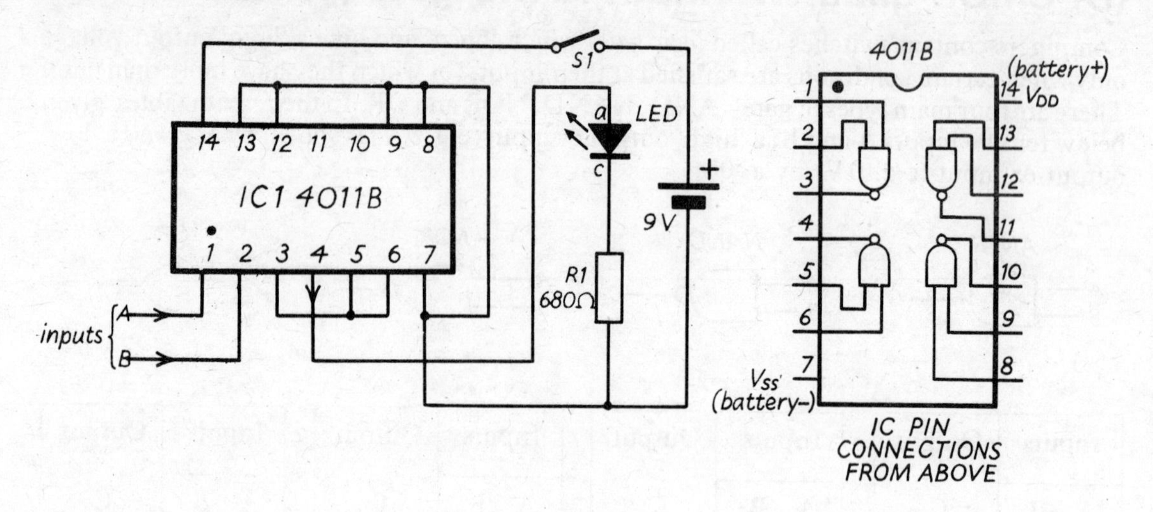

HOW TO BUILD IT

1 Identify pin 1 on the IC from the small dot or notch at one end of the case. Carefully push the IC into the circuit board in the position shown, taking care not to touch or bend the pins.

2 Insert wire links from the IC to the positive and negative rails and between other sockets on the board, as shown.

3 Insert *R*1.

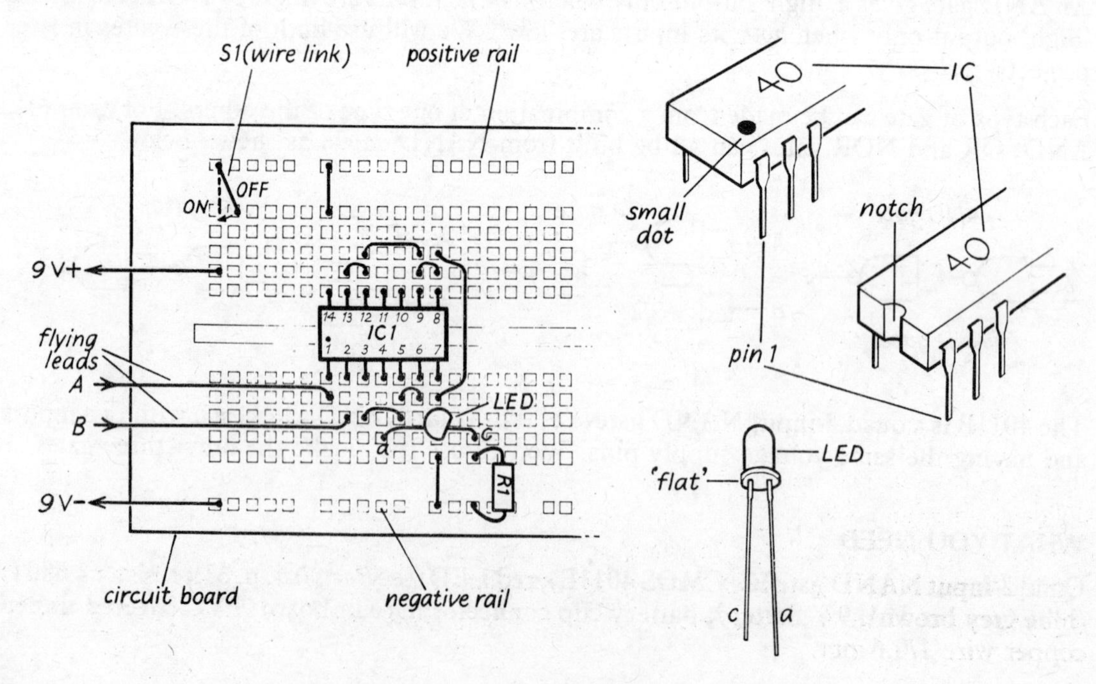

4 Insert the LED remembering that the cathode is nearest the 'flat' at the base of the case (or the cathode lead *c* may be shorter than the anode lead *a*).

5 CHECK THE CIRCUIT CAREFULLY.

6 Connect the battery with the correct polarity. The wire link acts as the battery on/off switch *S*1. Switch *S*1 on. Check the 'truth table' for an AND gate (page 19) by connecting

   a A and B to the positive rail so that both inputs are 'high': the output should be 'high' and so make the LED light up (if it doesn't it may be connected the wrong way round).

   b A to the positive rail ('high') and B to the negative rail ('low'): the output should be 'low' and the LED off,

   c A and B to the negative rail (both 'low'): the LED should remain off, showing that the output is again 'low',

   d A to the negative rail ('low') and B to the positive rail ('high'): the output should again be 'low'.

THINGS TO TRY

1 *NAND gate.* Remove the anode lead *a* of the LED from the socket in the circuit board connected to pin 4 of the IC and insert it in a socket connected to pin 3. The LED will then indicate the output level (at pin 3) of the NAND gate with inputs at pins 1 and 2. Check the 'truth table' for a NAND gate as you did for the AND gate.

2 *OR gate.* Using three of the NAND gates in the 4011B, connect up an OR gate as shown on page 19. Check its 'truth table'.

3 *NOR gate.* Using all four NAND gates in the 4011B, connect up a NOR gate as shown on page 19. Check its 'truth table'.

2 TWO-TONE DOOR BELL

After you have built this simple circuit you can 'ring' the changes and see how they affect the two-tone chimes.

WHAT YOU NEED

Dual astable multivibrator IC (556); resistors – 1 kΩ (brown black red), 22 kΩ (red red orange), 47 kΩ (yellow violet orange), two 100 kΩ (brown black yellow); electrolytic capacitors – two 1 μF, 10 μF, 100 μF; disc ceramic capacitors – two 0.01 μF, 0.1 μF; loudspeaker 2½ in, 25 to 80 Ω; miniature slide switch SPDT; 9 V battery; battery clip connector; circuit board; PVC-covered tinned copper wire 1/0.6 mm.

HERE IS THE CIRCUIT

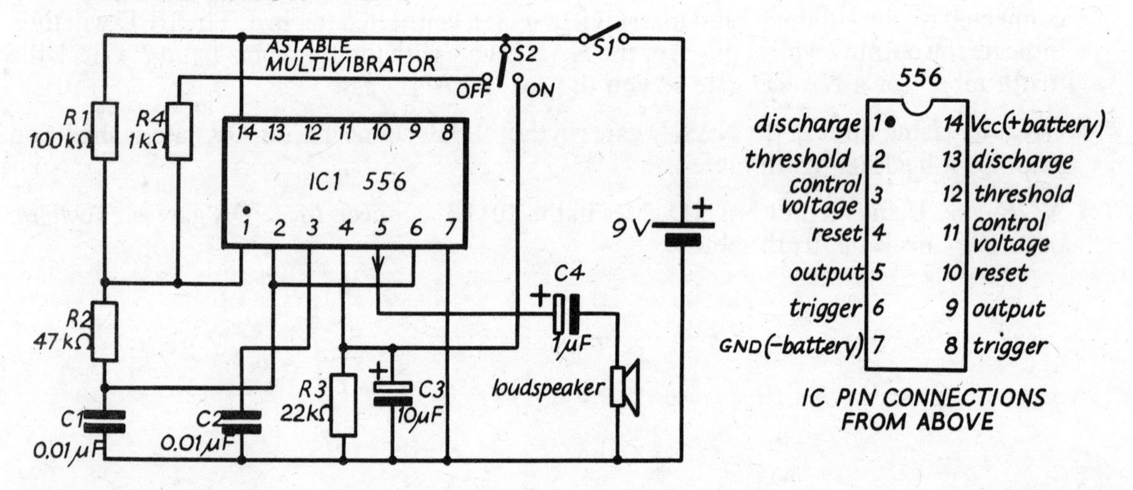

HOW TO BUILD IT

1 Identify pin 1 on the IC from the small dot or notch at one end of the case. Carefully push the IC into the circuit board in the position shown, taking care not to bend any of the pins.

2 Insert wire links from the IC to the positive and negative rails and between other sockets on the board, as shown.

3 Insert $R1, R2, R3, R4, C1, C2, C3, C4$ and $S2$. Be sure that the electrolytic capacitors $C3$ and $C4$ are connected the correct way round, the + end has a groove and the – end a black band round it (usually).

4 Connect the loudspeaker.

5 CHECK THE CIRCUIT CAREFULLY.

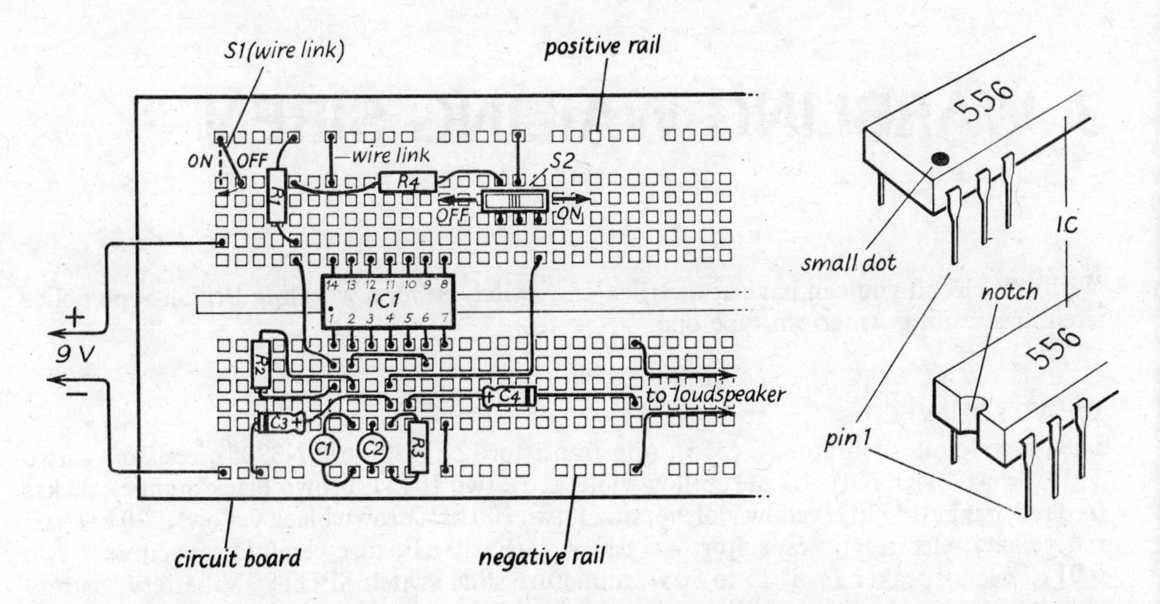

6 Connect the battery with the correct polarity. The wire link $S1$ acts as the battery on/off switch. Switch $S1$ on, then $S2$ *on and off*. A two-tone chime should be produced if all is well.

HOW IT WORKS

When switches $S1$ and $S2$ are both 'on', 'fast' square wave pulses from the astable produce a note in the loudspeaker. The pitch of this note depends on the values of $R1$, $R2$ and $C1$ (see page 11). At the same time $C3$ charges up. When $S2$ is switched 'off', the astable continues to produce output pulses (at pin 5) but faster than before because $R4$ (1 kΩ) is now in parallel with $R1$ (100 kΩ) and has effectively reduced the value of $R1$ (to less than 1 kΩ). A note of higher pitch is therefore produced. It stops when $C3$ has discharged sufficiently (through $R3$) for the voltage across it to be less than 0.7 V. The voltage at 'reset' (pin 4) is then too low for the astable to operate (see page 12).

THINGS TO TRY

1 Find out what happens to the two-tone chimes when

a $R1 = 100\,\text{k}\Omega$, $R2 = 100\,\text{k}\Omega$ and $C1 = 0.01\,\mu\text{F}$,

b $R1 = 100\,\text{k}\Omega$, $R2 = 47\,\text{k}\Omega$ and $C1 = 0.1\,\mu\text{F}$.

2 With $R1 = 100\,\text{k}\Omega$, $R2 = 47\,\text{k}\Omega$ and $C1 = 0.01\,\mu\text{F}$, try to guess what the result will be when you switch $S2$ 'on' and 'off' after

a changing $C3$ from 10 μF to 100 μF,

b making $C3 = 1\,\mu\text{F}$,

c removing $C3$ altogether from the circuit,

d making $C3 = 10\,\mu\text{F}$ but removing $R3$ from the circuit.

3 WARBLING-WAILING SIREN

With this circuit you can have at the flick of a switch either a warbling British-type police siren or a wailing American-type one.

WHAT YOU NEED

Dual astable multivibrator IC (556); npn transistor (ZTX300 or 2N3705); resistors – two 1 kΩ (brown black red), 4.7 kΩ (yellow violet red), two 10 kΩ (brown black orange), 22 kΩ (red red orange), 47 kΩ (yellow violet orange), two 100 kΩ (brown black yellow), 220 kΩ (red red yellow); electrolytic capacitors – 1 μF, 4.7 μF, 10 μF; disc ceramic capacitors – two 0.01 μF; loudspeaker 2½ in, 25 to 80 Ω; miniature slide switch SPDT; 9 V battery; battery clip connector; circuit board; PVC-covered tinned copper wire 1/0.6 mm.

HERE IS THE CIRCUIT

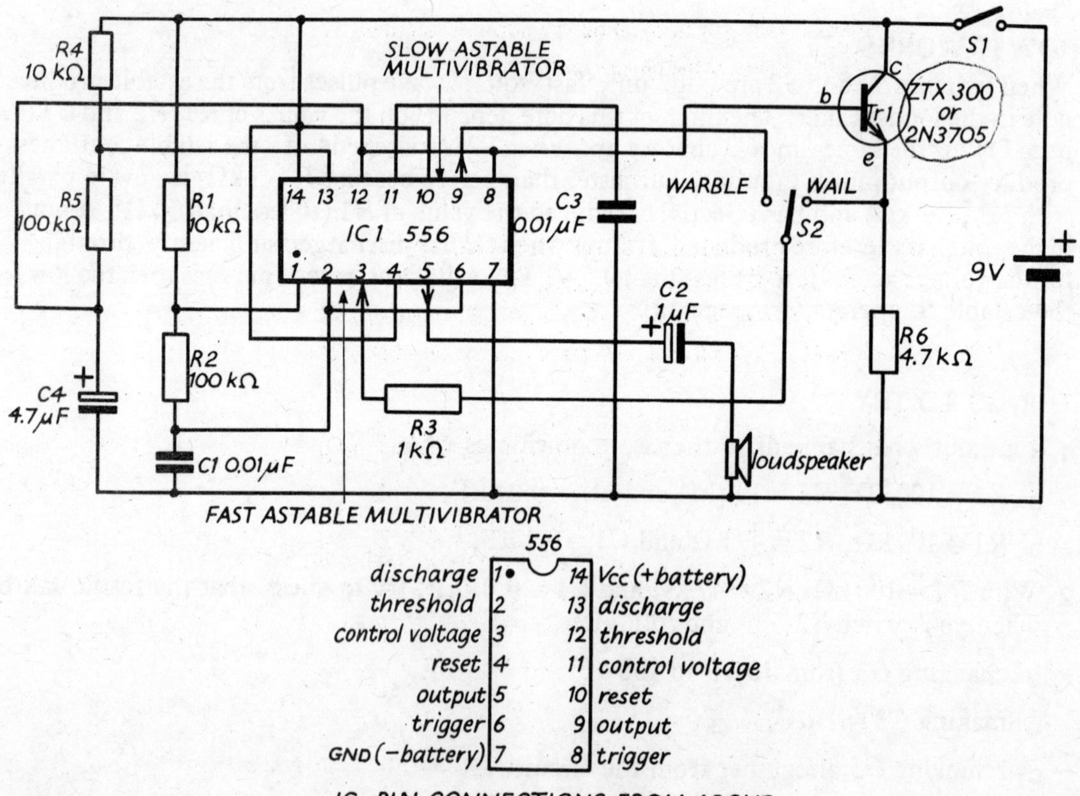

HOW TO BUILD IT

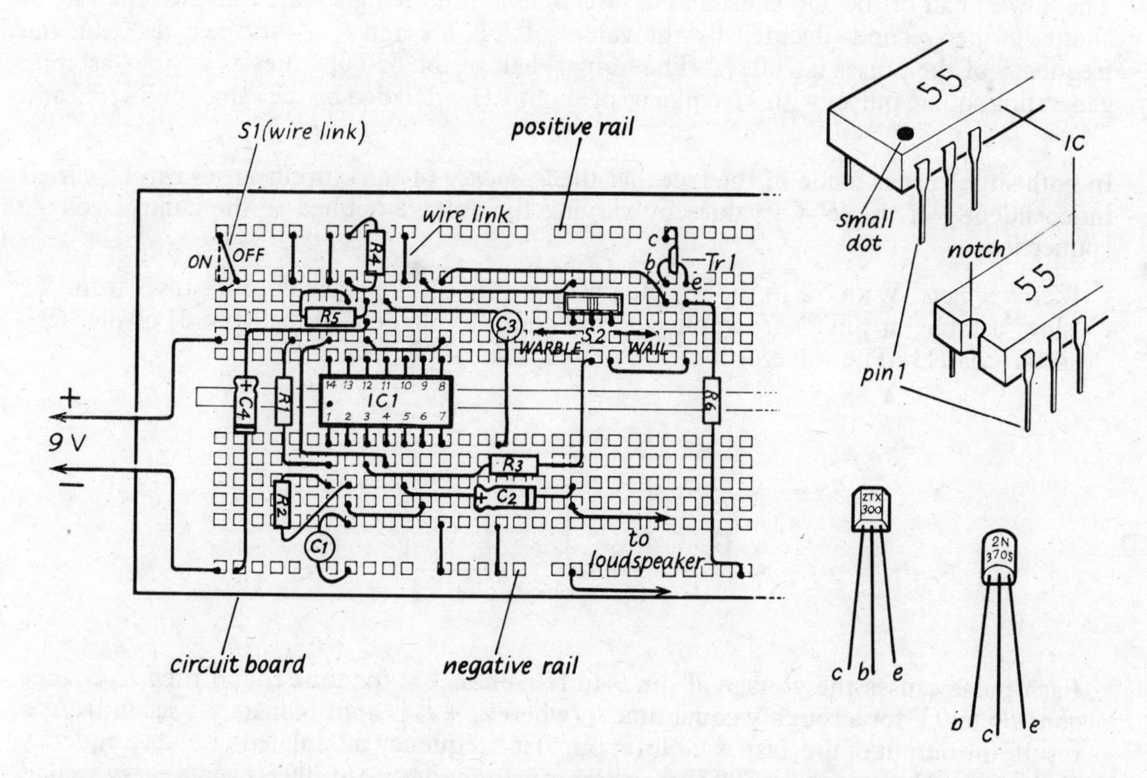

1 Identify pin 1 on the IC from the small dot or notch at one end of the case. Carefully push the IC into the circuit board in the position shown, taking care not to bend any of the pins.

2 Insert wire links from the IC to the positive and negative rails and between other sockets on the board, as shown.

3 Insert *R*1, *R*2, *R*3, *R*4, *R*5, *R*6, *C*1, *C*2, *C*3, *C*4 and *S*2. Be sure that the electrolytic capacitors *C*2 and *C*4 are connected the correct way round, the + end has a groove and the − end a black band round it (usually).

4 Identify the collector (c), base (b) and emitter (e) leads on the transistor. Insert it in the circuit board, making sure that you have it exactly as shown (for a ZTX300). Also connect the loudspeaker.

5 CHECK THE CIRCUIT CAREFULLY.

6 Connect the battery with the correct polarity (the wire link *S*1 acts as the battery on/off switch). Switch *S*1 on. With *S*2 switched to the left in the 'warble' position you should hear a British-type police siren; with *S*2 to the right in the 'wail' position you should hear an American-type one.

Note. If the noise of the siren disturbs anyone else you can listen to it yourself by disconnecting the loudspeaker and connecting one lead of a *crystal* earphone to one of the vacant sockets on the circuit board that goes to pin 5 on the IC and taking the other lead to the negative rail.

HOW IT WORKS

The 'lower' half of the 556 is used as a 'fast' astable producing square pulses at the rate of about 700 per second (decided by the values of $R1$, $R2$ and $C1$ – see page 11), i.e. the frequency of the pulses is 700 Hz. The 'upper' half of the 556 operates as a 'slow' astable, generating square pulses with a frequency of about 1 Hz (decided by the values of $R4$, $R5$ and $C4$).

In both sirens use is made of the fact that the *frequency* of the output pulses can be varied independently of the 'R–C' values by varying the voltage applied to the control voltage connection.

a *Warbling note.* With $S2$ in the warble position, the square wave output pulses from the 'slow' astable (at pin 9) are applied to the control voltage terminal (pin 3) of the 'fast' astable via $R3$. The waveform of the pulses is shown below.

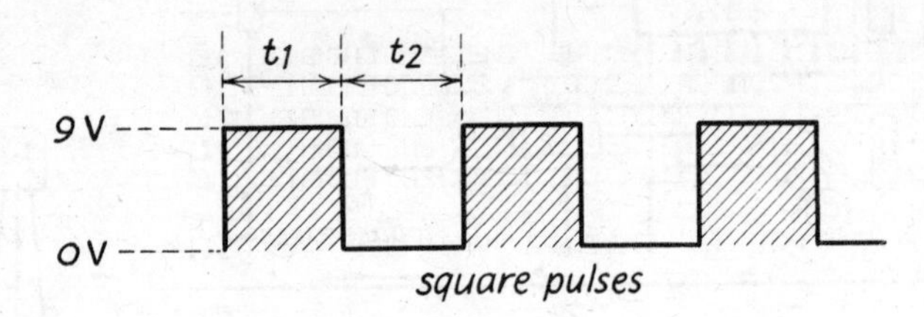

square pulses

Each pulse causes the voltage at pin 9 to remain at 9 V for time t_1 and then to change *abruptly* to 0 V for a roughly equal time t_2 (where $t_1 + t_2$ is approximately 1 second). As a result, the output of the 'fast' astable (at pin 5) is 'frequency modulated', i.e. its frequency changes *sharply* from about 700 Hz to a higher value and back to 700 Hz again every second or so. It produces this two-tone effect repeatedly so giving a warbling note.

b *Wailing note.* With $S2$ in the wail position, sawtooth-shaped pulses which occur across $C4$ (see page 12) are fed to the base of $Tr1$ and then via $S2$ and $R3$ to the control voltage terminal (pin 3) on the 'fast' astable.

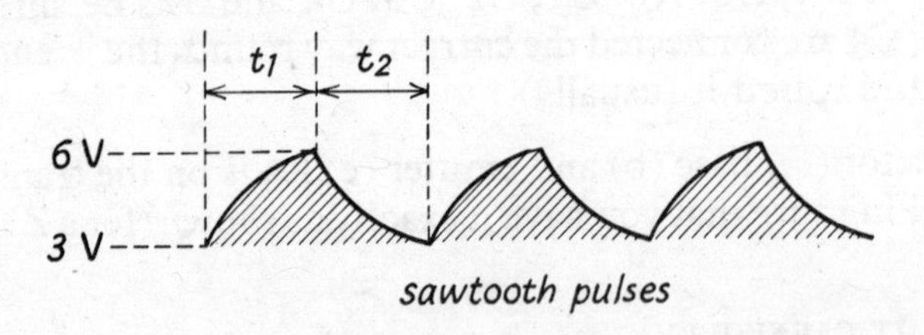

sawtooth pulses

Frequency modulation again occurs. This time, however, because of the shape of the sawtooth pulses, the control voltage is varied *slowly* and over a smaller range of values. As a result the frequency of the 'fast' astable output rises *slowly* from a low value to a high value in time t_1 and then falls again *slowly* to a low value in time t_2 (where $t_1 + t_2$ is approximately 1 second). The average value of the frequency about which the note rises and falls is 700 Hz. The wailing effect is repeated indefinitely.

$Tr1$ acts as an 'emitter follower'; its purpose is to ensure that there is maximum voltage transfer of the sawtooth pulses from the 'slow' astable to the 'fast' one.

THINGS TO TRY

Try to predict (and then check) the effect on both the warble and the wail of

a changing $C4$ from 4.7 μF to 10 μF,

b keeping $C4 = 10\,\mu$F and changing $R2$ from 100 kΩ to 220 kΩ,

c keeping $C4 = 10\,\mu$F and changing $R2$ from 220 kΩ to 47 kΩ,

d keeping $C4 = 10\,\mu$F and making $R2 = 100\,$kΩ, $R4 = 1\,$kΩ and $R5 = 22\,$kΩ.

Perhaps you will be able to find a combination which gives even more realistic warbles and wails than those produced by the original circuit.

4 TWO-OCTAVE ELECTRONIC ORGAN

You are unlikely to be invited to give concert performances with this 'organ' but at least you should manage to play some recognizable tunes.

WHAT YOU NEED

Dual astable multivibrator IC (556); resistors – 330 Ω (orange orange brown), 3.9 kΩ (orange white red), two 8.2 kΩ (grey red red), two 10 kΩ (brown black orange), 12 kΩ (brown red orange), four 15 kΩ (brown green orange), 18 kΩ (brown grey orange), two 22 kΩ (red red orange), two 27 kΩ (red violet orange), two 39 kΩ (orange white orange), two 47 kΩ (yellow violet orange), 100 kΩ (brown black yellow); electrolytic capacitor – 1 μF; disc ceramic capacitors – two 0.01 μF; loudspeaker 2½ in, 25 to 80 Ω; 9 V battery; battery clip connector; circuit board; PVC-covered tinned copper wire 1/0.6 mm.

HERE IS THE CIRCUIT

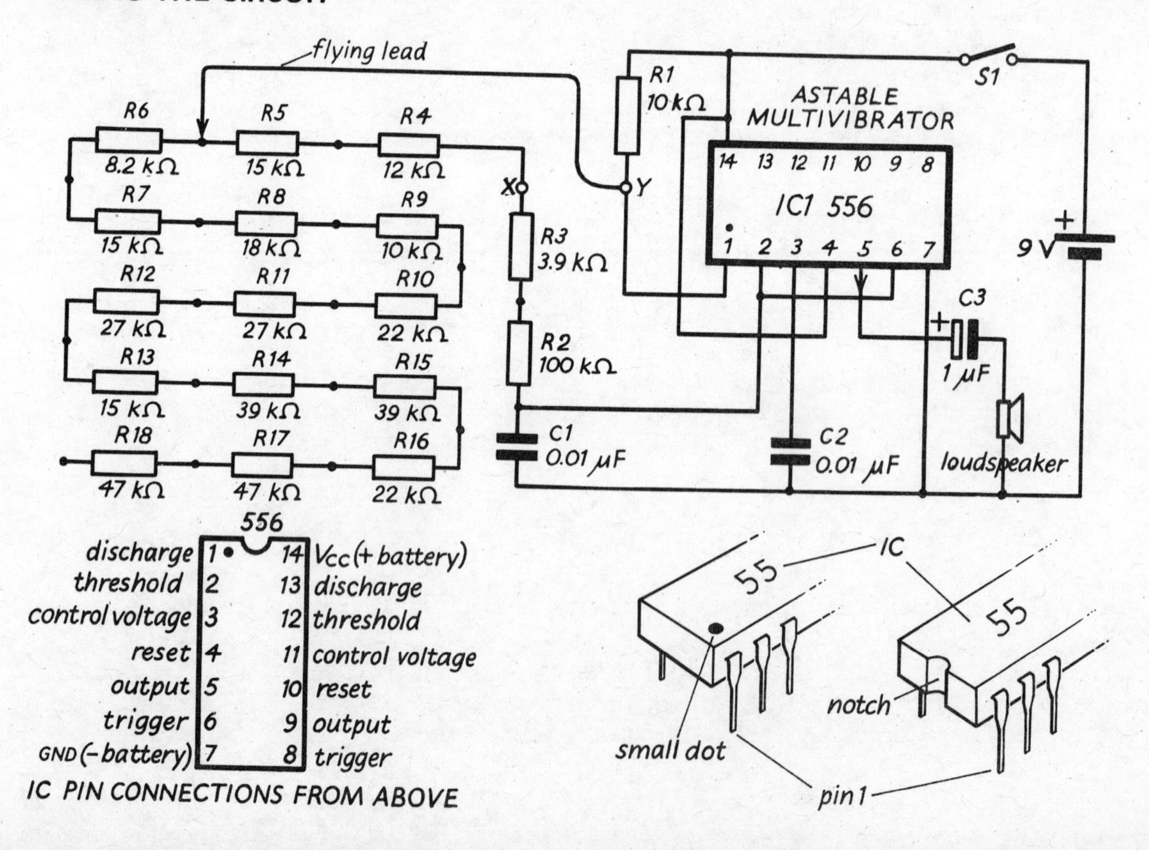

HOW TO BUILD IT

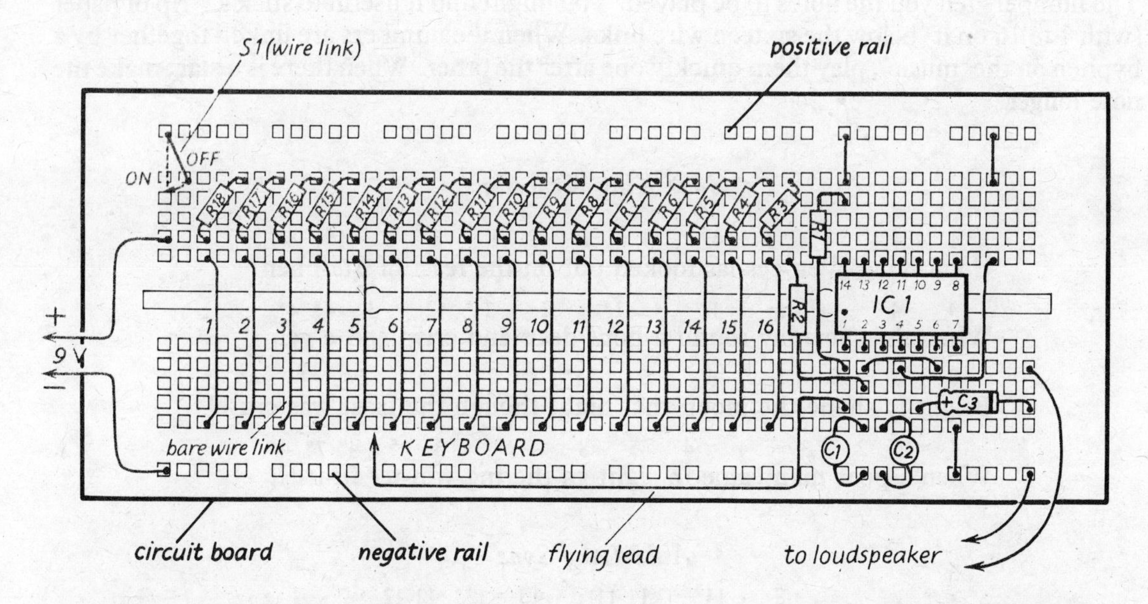

1 Identify pin 1 on the IC from the small dot or notch at one end of the case. Carefully push the IC into the circuit board in the position shown, taking care not to bend any of the pins.

2 Insert wire links from the IC to the positive and negative rails and between other sockets on the board, as shown.

3 Insert resistors $R1$ to $R18$, $C1$, $C2$ and $C3$. Be sure that the electrolytic capacitor $C3$ is connected the correct way round, the + end has a groove and the − end a black band round it (usually).

4 Connect the loudspeaker.

5 CHECK THE CIRCUIT CAREFULLY.

6 Connect the battery with the correct polarity (the wire link $S1$ acts as the battery on/off switch). Switch $S1$ on.

7 Touch the bare end of the flying lead on each of the sixteen wire links (the 'keyboard') in turn. They each give a different note. The lowest (note 1) is got from the left-hand end link (most resistance); the highest (note 16) from the end right-hand link. Doh is notes 4 and 11 of the scale.

With a bit of practice you should be able to play a few tunes. 'Music' for *Good King Wenceslas* and *Auld Lang Syne* is given on page 30.

HOW IT WORKS

One half of the 556 is used as an astable multivibrator. The frequency of the square wave pulses produced at its output (pin 5) and so also the pitch of the note emitted by the loudspeaker, depends on the position of the flying lead on the chain of fifteen resistors ($R4$ to $R18$). For the highest note only $R2$ and $R3$ are used.

'MUSIC' FOR SOME TUNES

The numbers tell you the notes to be played. You might find it useful to stick a strip of paper (with 1 to 16 on it) below the sixteen wire links. When the numbers are linked together by a hyphen on the 'music', play them quickly one after the other. When there is a star, make the note longer.

Good King Wenceslas

4 4 4 5 4 4 1* 2 1 2 3 4* 4*

Good King Wen-ces-las looked out on the feast of Ste-phen

4 4 4 5 4 4 1* 2 1 2 3 4* 4*

When the snow lay round a-bout deep and crisp and e-ven

8 7 6 5 6 5 4* 2 1 2 3 4* 4*

Bright-ly shone the moon that night though the frost was cru-el

1 1 2 3 4 4 5* 8 7 6 5 4* 7* 4*

When a poor man came in sight ga-thering win-ter fu-u-el.

Auld Lang Syne

8 11* 11-11 13 12* 11-12

Should auld ac-quain-tance be for-got

13 11* 11-13 15 16*

And ne-ver brought to mind

16 15* 13-13 11 12* 11-12

Should auld ac-quain-tance be for-got

13 11* 9-9 8 11*

And days of auld lang syne

16 15* 13-13 11 12* 11-12

For au-ld la-ng syne my dear

16 15* 13-13 15* 16*

For au-ld la-ng syne

16 15* 13-13 11 12 11-12

We'll tak' a cup o' kind-ness yet

13 11* 9-9 8 11*

For au-ld la-ng syne.

THINGS TO TRY

1 *Reducing loudspeaker volume.* Connect a 330 Ω resistor in series with *C*3 and the loudspeaker, as shown.

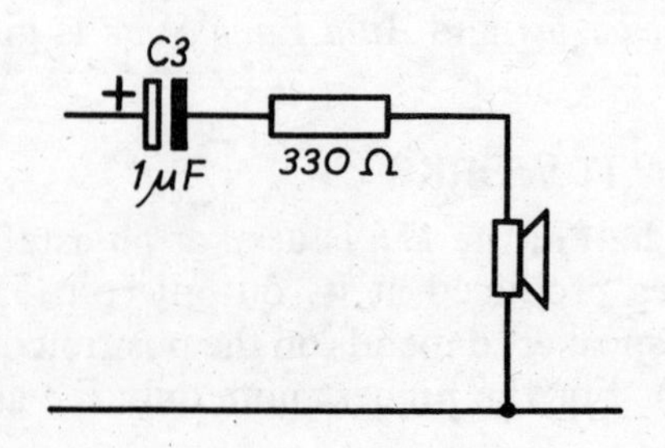

2 *Semitones*. You can make a new scale with Doh as notes 1, 8 and 15. To do this move the lower end of the wire link for note 8 up one hole. Put one end of a 15 kΩ resistor in the hole that is now free and the other end in the middle hole *between* notes 7 and 8, as shown.

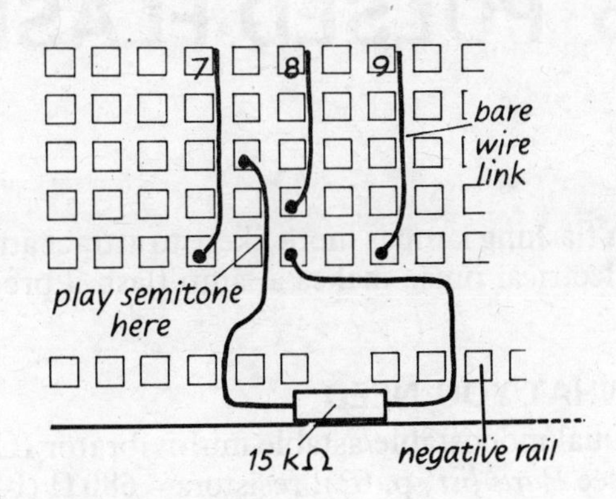

Play the new note on the resistor wire itself. The scale now goes: 1, 2, 3, 4, 5, 6, semitone, 8. To extend your scale, do the same between notes 14 and 15, using an 8.2 kΩ resistor. Try playing *Good King Wenceslas* starting on note 8.

You can add other semitones in the same way and play more difficult tunes. Use a resistor of about half the value of the one between the two notes where the semitone is to go.

3 *Construct, and connect to the circuit board, a better keyboard.*

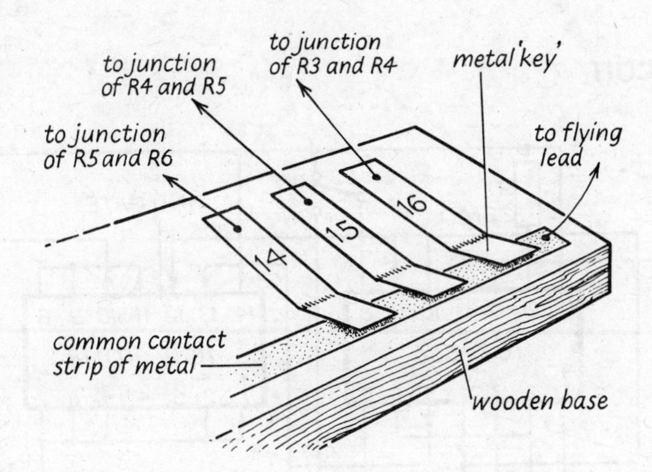

For example, you could make the keys from narrow strips of thin brass sheet (or tinplate) each of which completes the astable circuit (through one or more of the resistors in the chain) when it is pressed onto a common contact strip.

5 PULSED FLASHING LAMP

A flashing lamp is more likely to attract attention than one that is always on. In this circuit an electrical pulse makes a lamp flash a predetermined number of times.

WHAT YOU NEED

Dual monostable/astable multivibrator IC (556); NAND gate IC (CMOS 4011B); red LED (see *Parts list*, p. 63); resistors – 680 Ω (blue grey brown), 1 kΩ (brown black red), 220 kΩ (red red yellow), two 2.2 MΩ (red red green); electrolytic capacitors – 1 μF, 10 μF; disc ceramic capacitors – 0.01 μF, 0.1 μF; miniature slide switch SPDT; 9 V battery; battery clip connector; circuit board; PVC-covered tinned copper wire 1/0.6 mm.

HERE IS THE CIRCUIT

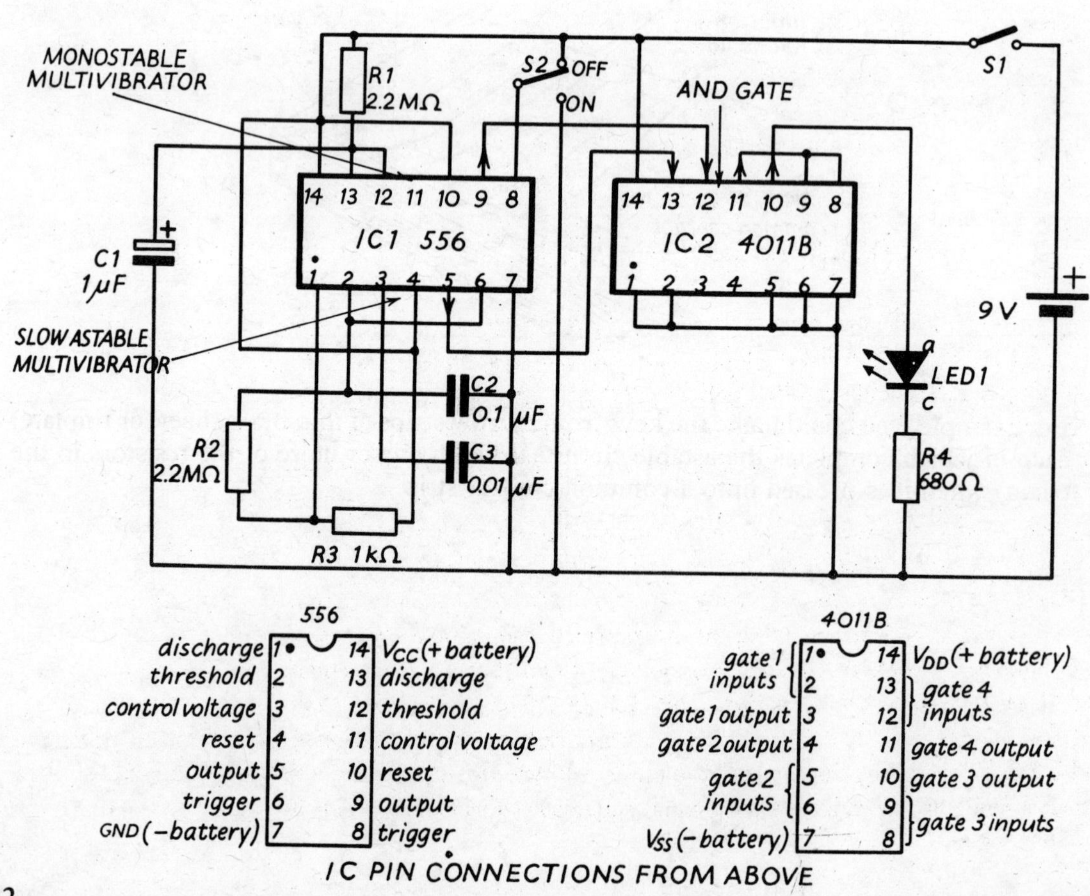

IC PIN CONNECTIONS FROM ABOVE

HOW TO BUILD IT

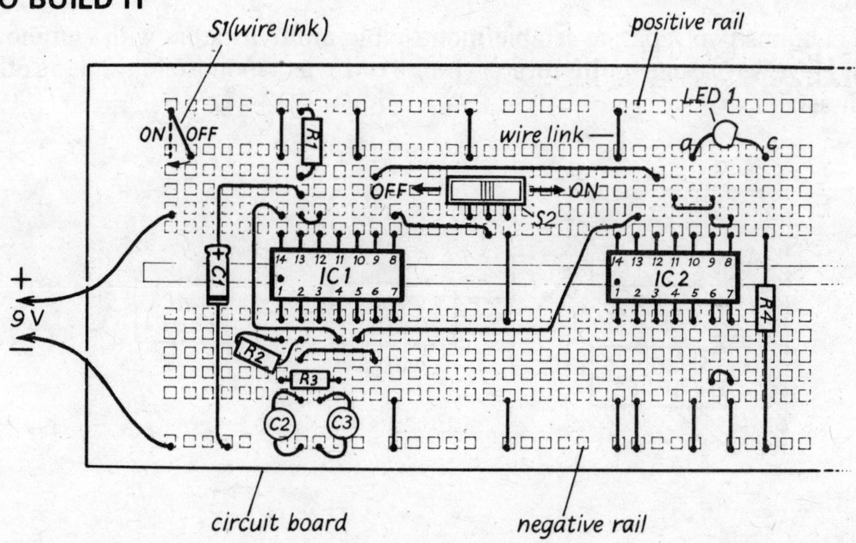

1 Identify pin 1 on the ICs from the small dot or notch at one end of the case. Carefully push each IC into the circuit board in the position shown, taking care not to bend any of the pins nor to touch those on the CMOS IC.

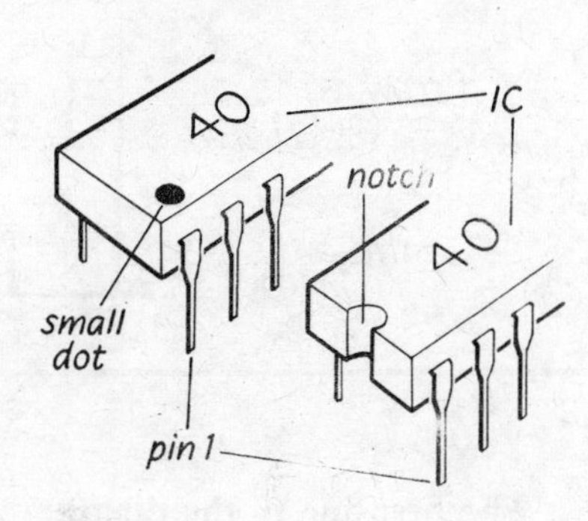

2 Insert wire links from each IC to the positive and negative rails and between other sockets, as shown.

3 Insert $R1, R2, R3, R4, C1, C2, C3$ and $S2$. Be sure that $C1$ is connected the correct way round, the + end (it has a groove) should go to the bottom of $R1$.

4 Insert the LED, remembering that the cathode is nearest the 'flat' at the base of the case (or the cathode lead *c* may be shorter than the anode lead *a*).

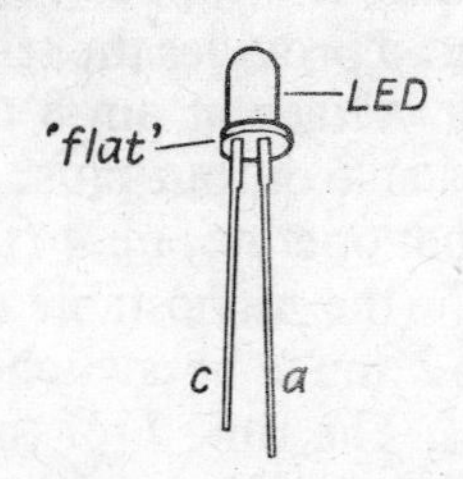

5 CHECK THE CIRCUIT CAREFULLY.

6 Connect the battery with the correct polarity (the wire link $S1$ acts as the battery on/off switch). Switch $S1$ on, then $S2$ on *and* off. (Switch $S2$ off immediately after you have switched it on, otherwise leaving $S2$ on produces continuous flashing.) The LED should flash six or so times; if it doesn't it may be connected the wrong way round. Every time you switch $S2$ on *and* off, the LED is 'pulsed' to produce a set of flashes.

HOW IT WORKS

The 556 IC contains two separate astable/monostable multivibrators with common supply voltage pins. Here we use one multivibrator (pins 8 to 13) as a monostable and the other (pins 1 to 6) as an astable.

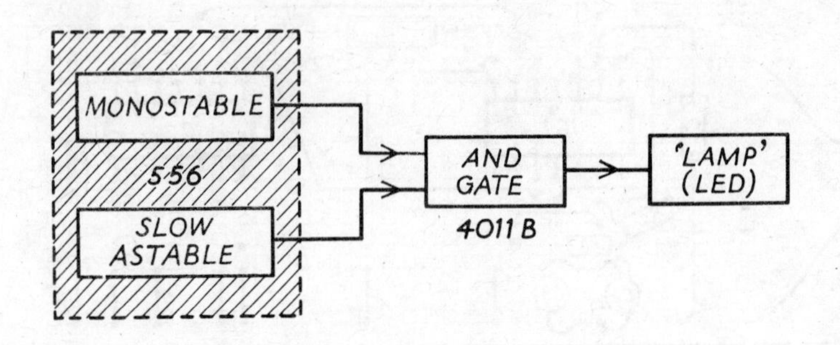

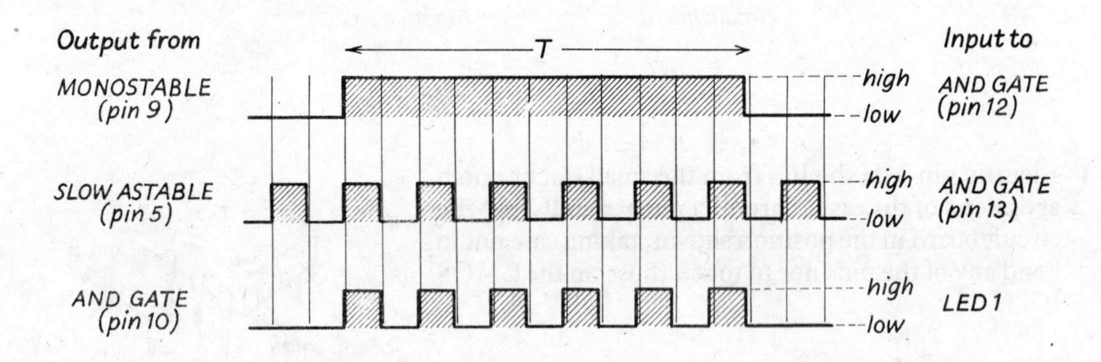

The first line in the diagram above shows the monostable output pulse (at pin 9) which is produced when the monostable is triggered (at pin 8) by the *negative* edge of a pulse caused by switching $S2$ on and off. With $S1$ on and $S2$ off, pin 8 is at 9 V, but on switching $S2$ on as well, pin 8 becomes connected to 0 V. The negative edge AB so produced provides the triggering for the monostable. The trigger voltage at pin 8 must return to + 9 V before the monostable output pulse ends, otherwise the monostable does not operate, i.e. t (the triggering time) must be less than T (the monostable output pulse time). This means that $S2$ must be switched on and off quickly for this circuit. The time T of the monostable pulse depends on the values of $R1$ and $C1$ (see page 13). The pulse is fed to one input (pin 12) of the AND gate.

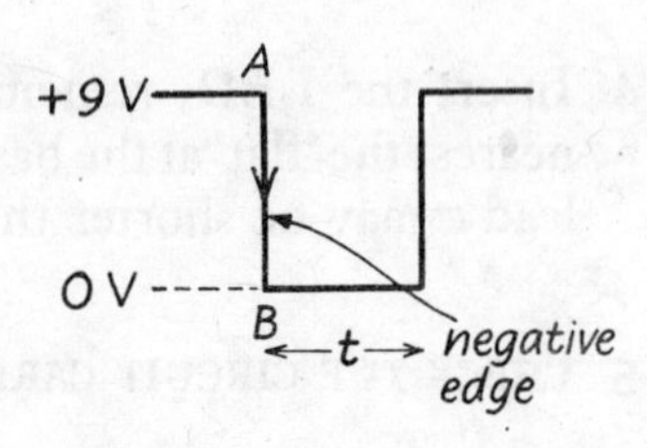

The second line shows the astable output pulses produced at pin 5 when $S1$ is closed. The rate at which they occur depends on the values of $R2$, $C2$ and $R3$ (see page 11) but it is quite slow in this circuit so that the flash of light produced on the LED by *each* pulse can be seen. They supply the second input to the AND gate (pin 13).

An AND gate 'opens' and gives a 'high' output only when *both* its inputs are 'high' (see page 19). From the first and second lines on the diagram you can see why the third line gives the output of the AND gate (at pin 10) – it is 'high' when both inputs are 'high'. In the diagram this occurs six times during the monostable pulse and so the LED flashes six times. The AND gate is obtained by connecting two of the four NAND gates in the 4011B (see page 19), i.e. gates 3 and 4.

THINGS TO TRY

1. *Effect of $C1$.* Change $C1$ from 1 μF to 10 μF. The monostable output pulse now lasts for 20 seconds or so (compared with 2 seconds previously). This allows the LED to flash for longer.
2. *Effect of $R1$.* Keep $C1 = 10\ \mu$F but change $R1$ from 2.2 MΩ to 220 kΩ. The monostable output pulse returns to its former value of about 2 seconds – as does the time for which the LED flashes.
3. *Effect of $R2$.* Keep $C1 = 10\ \mu$F but make $R1 = 2.2$ MΩ again and replace $R2$ (2.2 MΩ) by a 10 MΩ resistor. The LED flashes for about 20 seconds (i.e. the time of the monostable pulse) but there are fewer flashes because the larger value of $R2$ has reduced the number of astable pulses per second. How else could you change the astable pulse rate?

6 LIGHT-OPERATED ALARM

This alarm gives either a visual or an audible warning, the choice is yours. It springs into action when the intensity of the light falling on a photocell changes. For example, at dusk when there is a decrease, it could switch on automatically a flashing lamp to give warning of a road hazard. Or it could rouse you at the crack of dawn (as the light intensity increases) with a persistent, gradually rising 'howl'. It could also be used as a burglar alarm.

WHAT YOU NEED

Astable multivibrator IC (CMOS 4047B); photocell (e.g. ORP12); two npn transistors (ZTX300 or 2N3705); red LED (see *Parts list*, p. 63); resistors – 100 Ω (brown black brown), 680 Ω (blue grey brown), 4.7 kΩ (yellow violet red), two 10 kΩ (brown black orange), 15 kΩ (brown green orange), 22 kΩ (red red orange), 100 kΩ (brown black yellow), 2.2 MΩ (red red green); disc ceramic capacitors – 0.01 μF, 0.1 μF; loudspeaker 2½ in, 25 to 80 Ω; miniature slide switch SPDT; 9 V battery; battery clip connector; circuit board; PVC-covered tinned copper wire 1/0.6 mm.

HERE IS THE CIRCUIT

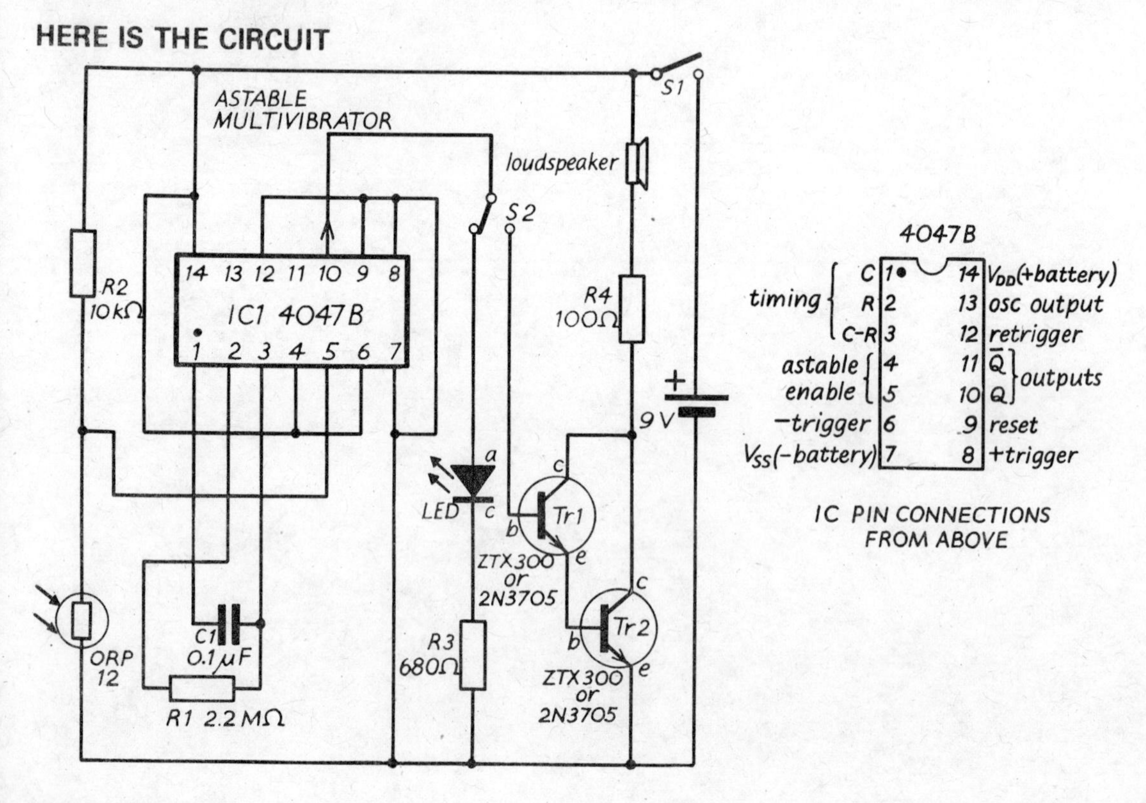

HOW TO BUILD IT

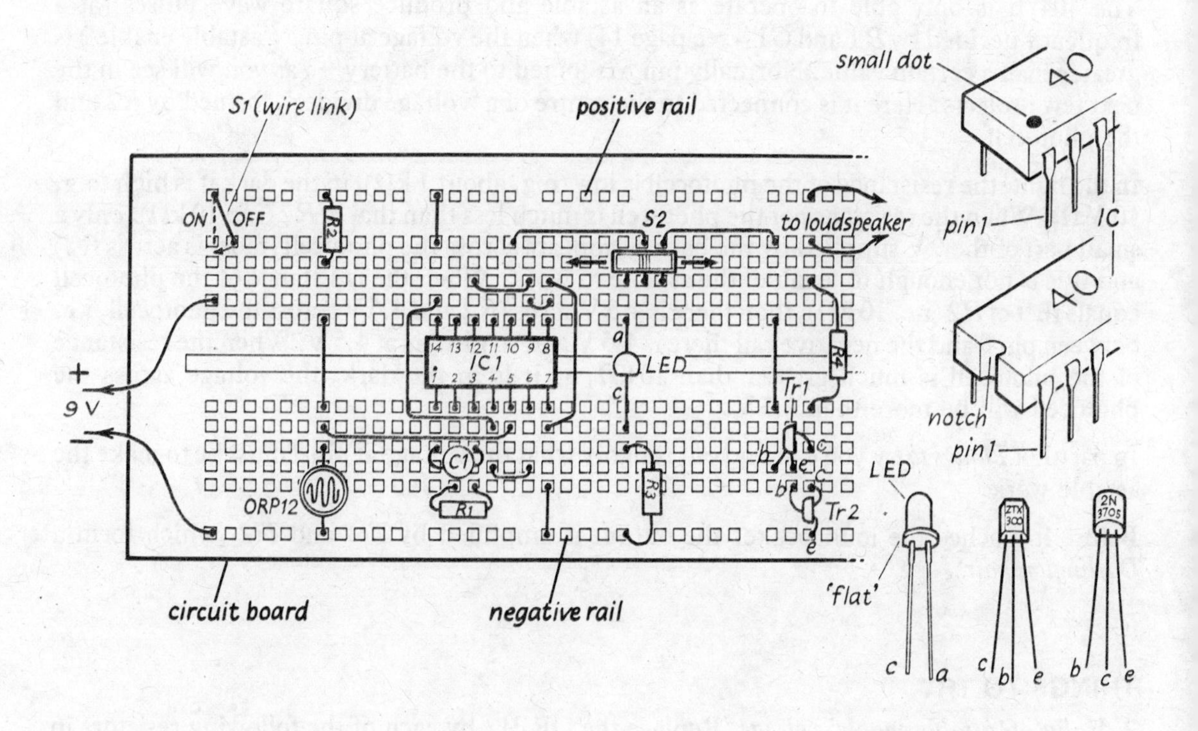

1 Identify pin 1 on the IC from the small dot or notch at one end of the case. Carefully push the IC into the circuit board in the position shown, taking care not to bend or touch any of the pins.

2 Insert wire links from the IC to the positive and negative rails and between other sockets, as shown.

3 Insert $R1$, $R2$, $R3$, $R4$, $C1$, $S2$ and the photocell (e.g. ORP12).

4 Insert the LED remembering that the cathode is nearest the 'flat' at the base of the case (or the cathode lead *c* may be shorter than the anode lead *a*).

5 Identify the collector (c), base (b) and emitter (e) leads on the transistors. Insert them in the circuit board making sure you have them exactly as shown (for ZTX300s) and that the leads are not touching one another. Also connect the loudspeaker.

6 CHECK THE CIRCUIT CAREFULLY.

7 Connect the battery with the correct polarity (the wire link $S1$ acts as the battery on/off switch). Switch $S1$ on. IN THE DARK (or with the photocell covered by a handkerchief), with $S2$ switched to the left, the LED should flash slowly; with $S2$ switched to the right slow 'clicks' should come from the loudspeaker.

8 If you now change $C1$ from 0.1 μF to 0.01 μF and $R1$ from 2.2 MΩ to 100 kΩ, then, IN THE DARK, the LED should appear to be on all the time (in fact it is flashing too rapidly for the eye to follow it) and the loudspeaker emits a low note.

HOW IT WORKS

The 4047B is only able to operate as an astable and produce square wave pulses (at a frequency decided by $R1$ and $C1$ – see page 14) when the voltage at pin 5 ('astable enable') is greater than a certain value. Normally pin 5 is joined to the battery +, as you will see in the next few projects. Here it is connected to the centre of a 'voltage divider', formed by $R2$ and the photocell.

In the light, the resistance of the photocell is low (e.g. about 1 kΩ), in the dark it is high (e.g. 10 MΩ). When the resistance of the photocell is much less than that of $R2$ (i.e. 10 kΩ), only a small part of the 9 V supplied by the battery appears across the photocell (most is across $R2$) and this is not enough to 'enable' the astable to work. When the resistance of the photocell equals that of $R2$ (i.e. 10 kΩ), then there is 4.5 V across $R2$ and 4.5 V across the photocell, i.e. between pin 5 and the negative rail there is 4.5 V and so pin 5 is at 4.5 V. When the resistance of the photocell is much greater than 10 kΩ, as it is in the dark, the voltage across the photocell will be more than 4.5 V.

In part 1 of *Things to try* you can find out roughly what the voltage at pin 5 must be to make the astable work.

Before it reaches the loudspeaker the output is amplified by $Tr1$ and $Tr2$ (which form a *Darlington pair*).

THINGS TO TRY

1 *Value of 'astable enable' voltage*. Replace the ORP12 by each of the following resistors in turn, in this order – 4.7 kΩ, 10 kΩ, 15 kΩ, 22 kΩ. What is the *lowest* value which sets the alarm working? Knowing that a voltage divides in the ratio of the resistances across which it is applied, estimate what the minimum voltage must be at pin 5 for the astable to operate.

2 *Burglar (or early-morning) alarm*. Interchange $R2$ (10 kΩ) and the photocell so that they are as shown in the diagram below. You should now find that IN THE LIGHT, the alarm is on all the time but that it goes off in the dark or when you cover the photocell. Connected in this way, it could be used to detect an unwelcome 'guest' shining a torch in your bedroom at night or, if getting up in the morning is a problem, it would rouse you at first light!

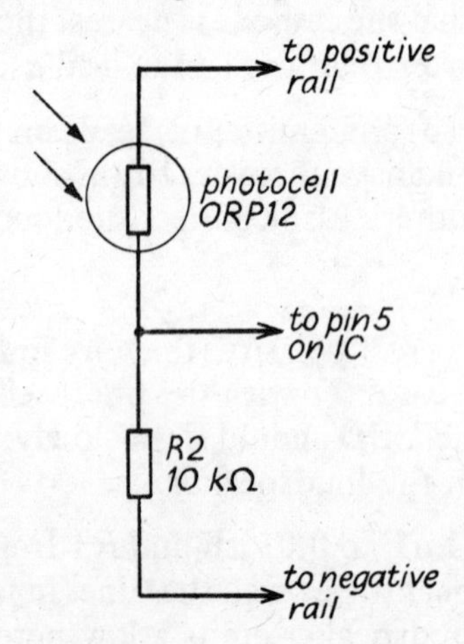

3 *Alternative version of burglar alarm*. Return $R2$ (10 kΩ) to its original position in the circuit, change $C1$ from 0.01 μF to 0.1 μF, remove $R1$ (100 kΩ) and replace it by the photocell. These changes are shown below. With light falling on the photocell the alarm will be on.

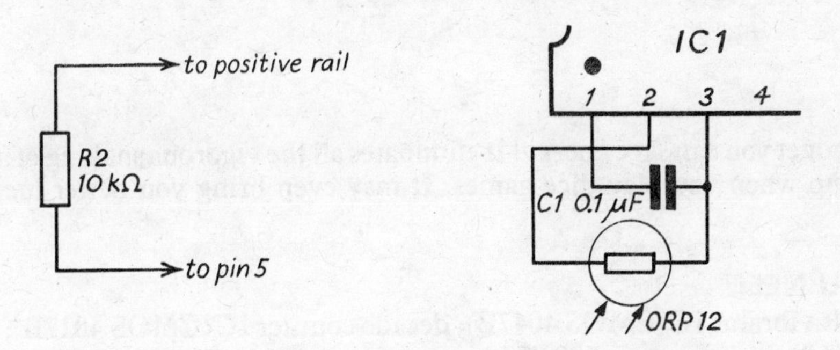

Now cover the photocell *completely* with your handkerchief until it stops working. If there is still the odd 'flash' from the LED or 'click' from the loudspeaker, then some light is reaching the photocell. Try to cut it all off. When you slowly remove your handkerchief, the LED will start flashing, slowly at first then faster till it appears to be on all the time and the loudspeaker will produce a series of clicks which soon become a continuous note.

7 ELECTRONIC DICE

With this gadget you can save energy! It eliminates all the vigorous shaking of a small wooden cube in a cup when you play dice games. It may even bring you better luck.

WHAT YOU NEED

Astable multivibrator IC (CMOS 4047B); decade counter IC (CMOS 4017B); six LEDs (see *Parts list*, p. 63); resistors – six 680 Ω (blue grey brown), 10 kΩ (brown black orange), 10 MΩ (brown black blue); disc ceramic capacitor – 0.01 μF; 9 V battery; battery clip connector; circuit board; miniature slide switch SPDT; push button on/off switch (optional); PVC-covered tinned copper wire 1/0.6 mm; plastic sleeving 1 mm bore (optional).

HERE IS THE CIRCUIT

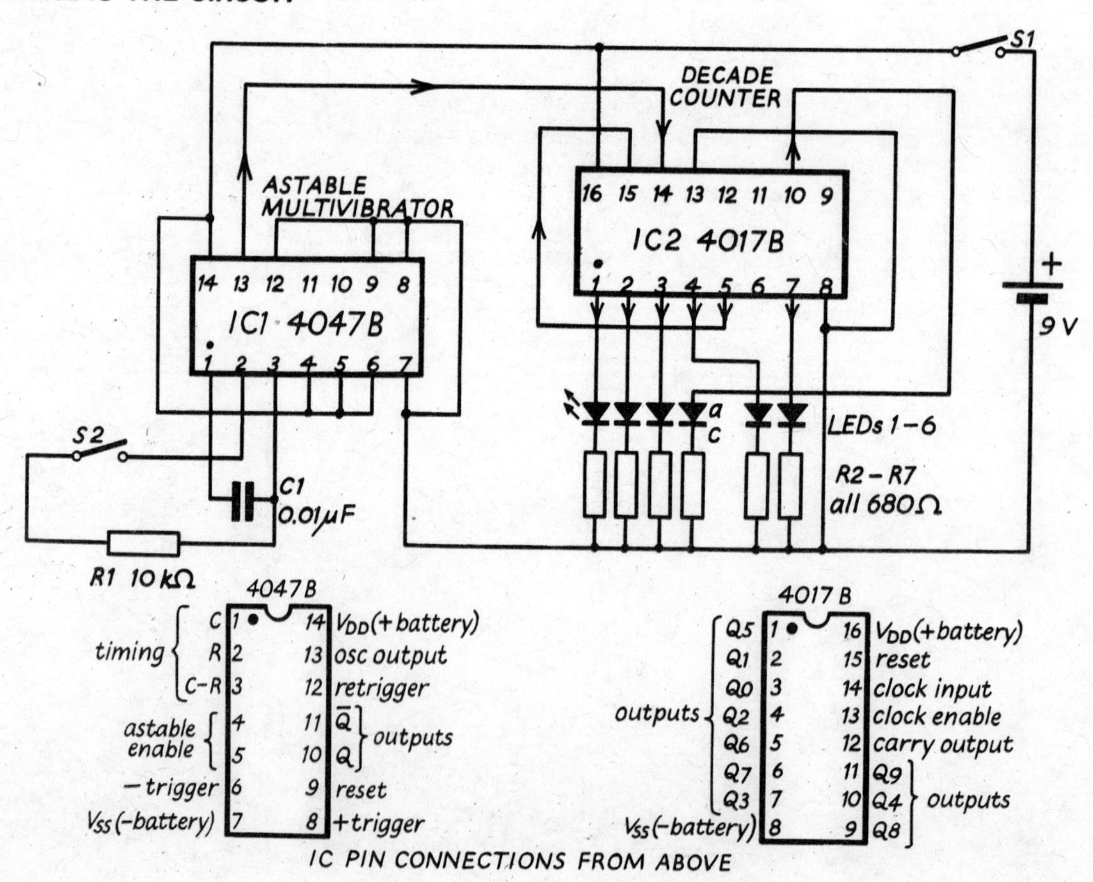

HOW TO BUILD IT

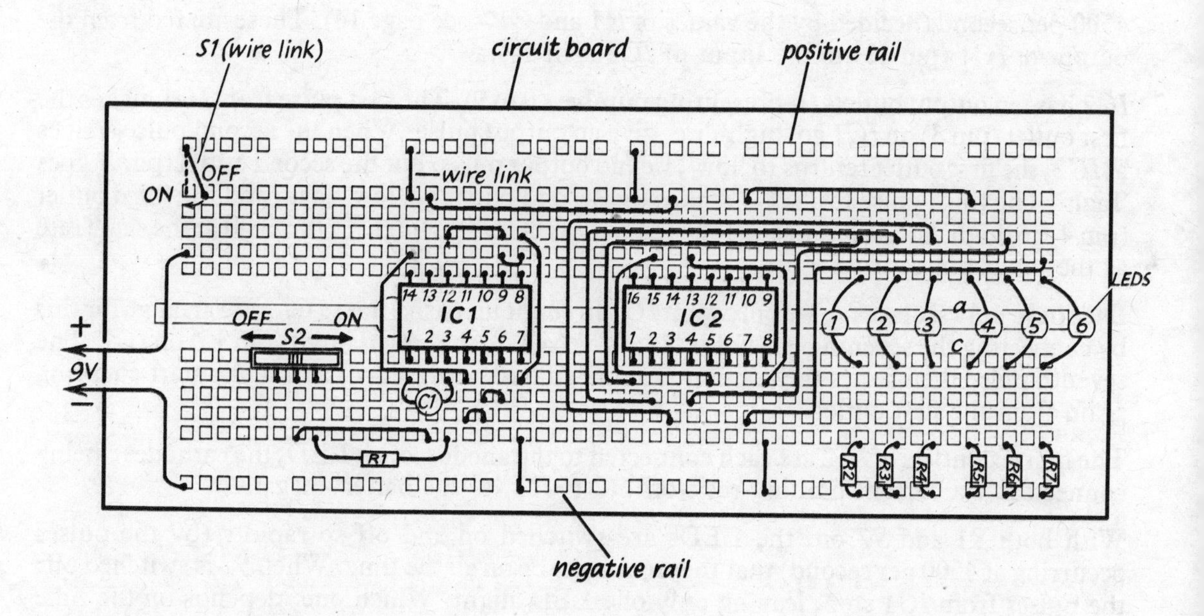

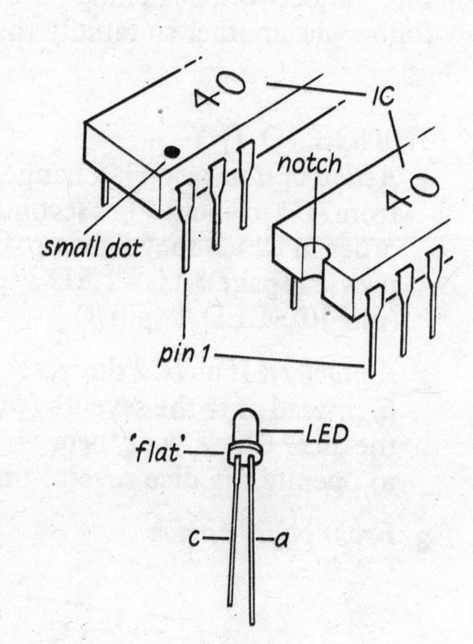

1 Identify pin 1 on the ICs from the small dot or notch at the end of the case. Carefully push each IC into the circuit in the positions shown, taking care not to bend or touch any of the pins in the process.

2 Insert wire links from each IC to the positive and negative rails and between other sockets, as shown.

3 Insert $R1$, $C1$, $R2$, $R3$, $R4$, $R5$, $R6$, $R7$ and $S2$.

4 Insert the six LEDs remembering that the cathode is next to the 'flat' at the bottom of the plastic case (or the cathode lead c may be shorter than the anode lead a).

5 CHECK THE CIRCUIT CAREFULLY.

6 Connect the battery with the correct polarity (the small wire link will act as the battery switch $S1$). Switch $S1$ on, then $S2$. All the LEDs should light; if any don't, they may be connected the wrong way round. Switch $S2$ off, only one LED should be alight and if you give each LED a number this will be the number you have 'thrown' with the dice. You can then repeat the process as often as you like simply by switching $S2$ on and off.

HOW IT WORKS

When $S1$ and $S2$ are both switched on, $IC1$ produces square pulses at the steady rate of about 4500 per second (decided by the values of $R1$ and $C1$ – see page 14). These are fed from the output of $IC1$ (pin 13) to the input of $IC2$ (pin 14).

$IC2$ has ten output outlets (representing numbers 0 to 9). The first pulse from $IC1$ makes the first outlet (pin 3) on $IC1$ go 'high', i.e. give an output pulse. When the second pulse arrives at $IC2$, the first outlet returns to 'low', i.e. no output pulse, but the second outlet (pin 2) goes 'high'. The third pulse from $IC1$ makes the second outlet go 'low' again and the third outlet (pin 4) go 'high' and so on. Each of the ten outlets of $IC2$ go 'high' in turn and at the same rate as the input pulses from $IC1$.

Dice have only six faces so we only want $IC2$ to count up to six, not to ten. We arrange for this by connecting the seventh outlet of $IC2$ (pin 5) back to the 'reset' input (pin 15) on $IC2$. The seventh pulse then resets the counter (so that all outlets are 'low') and it can start counting again from the first outlet.

The first six outlets of $IC2$ are each connected to the anode *a* of an LED, the cathodes *c* being connected, via current-limiting resistors $R2$ to $R7$, to the battery negative.

With both $S1$ and $S2$ on, the LEDs are switched on and off so rapidly (by the pulses occurring at 4500 per second) that they *appear* to be on all the time. When $S2$ is switched off, the pulses from $IC1$ stop, leaving only one LED alight. Which one, depends on the time interval between switching $S2$ on and off. This almost inevitably varies because the pulses follow one another so rapidly that you cannot time the switching of $S2$ accurately enough.

THINGS TO TRY

1 *Action in slow motion.* Change $R1$ from 10 kΩ to 10 MΩ. This will slow down the pulses from $IC1$ to about 4 per second and allow you to see each LED coming on in turn. Note the order in which they light up – taking the numbering on the circuit board diagram (top of previous page): it is – LED 3 (pin 3), LED 2 (pin 2), LED 5 (pin 4), LED 6 (pin 7), LED 4 (pin 10), LED 1 (pin 1).

2 *A problem.* If on $IC2$ the 'reset' connection (pin 15) is taken to the eighth output outlet (pin 6) instead of to the seventh (pin 5), can you work out how this will affect the behaviour of the dice? Check your theory by changing the connection. To make the check you will have to operate the dice several times.

3 *Front panel version.*

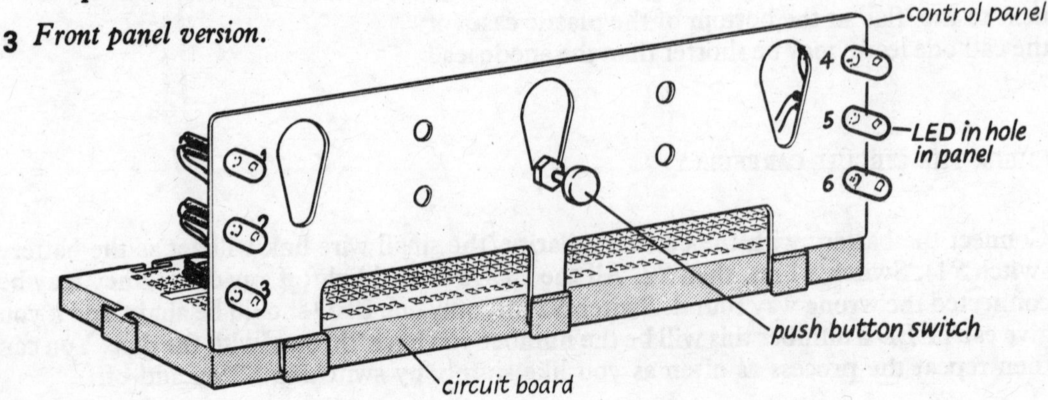

If you want to keep your electronic dice assembled for some time, you can give it a more professional look by securing the six LEDs in the holes in the control panel, mounted on the circuit board.

Wire the LEDs into the circuit by making connections to them using 1 mm bore plastic sleeving (see page 8). Also replace the slide switch *S*2 by a miniature push button on/off switch (an optional extra), mounting it on the panel as well.

To operate the dice you press the push button switch (so lighting all the LEDs) and then release it (leaving one LED alight).

4 *American traffic signals.* The order of the signals is RED, GREEN, AMBER, RED. You can make a red LED, a green LED and a yellow LED flash in this sequence if you rearrange the connections to the decade counter as shown below with R_1 as 10 MΩ. Only the Q_0, Q_1 and Q_2 outputs from pins 3, 2 and 4 respectively are connected to LEDs and the Q_3 output from pin 7 is returned to 'reset' (pin 15) since we just need to count up to three. Try it out.

Modify the circuit so that there is a pause (no LEDs on) between one LED going off and the next one coming on.

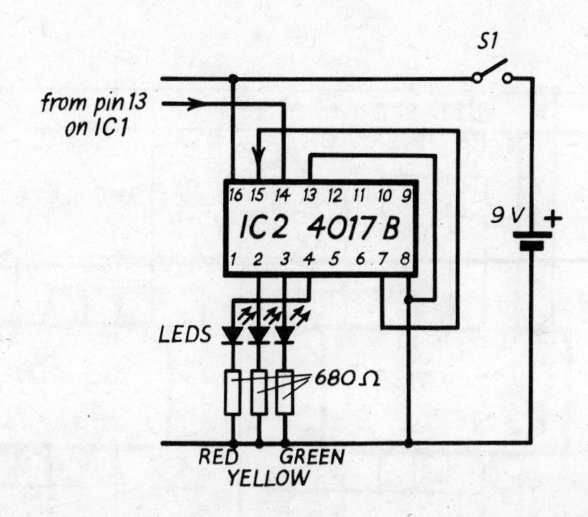

8 TRAFFIC LIGHTS

When you have completed this project successfully you should never have any trouble remembering whether it is RED (stop) or GREEN (go) that follows AMBER when amber has been on by itself in British traffic lights.

WHAT YOU NEED

Astable multivibrator IC (CMOS 4047B); bistable multivibrator IC (CMOS 4013B); NAND gate IC (CMOS 4011B); two npn transistors (ZTX300 or 2N3705); three LEDs (red yellow green) (see *Parts list*, p. 63); resistors – two 680 Ω (blue grey brown), 4.7 kΩ (yellow violet red), 2.2 MΩ (red red green), 10 MΩ (brown black blue); disc ceramic capacitor – 0.1 μF; 9 V battery; battery clip connector; circuit board; PVC-covered tinned copper wire 1/0.6 mm.

HERE IS THE CIRCUIT

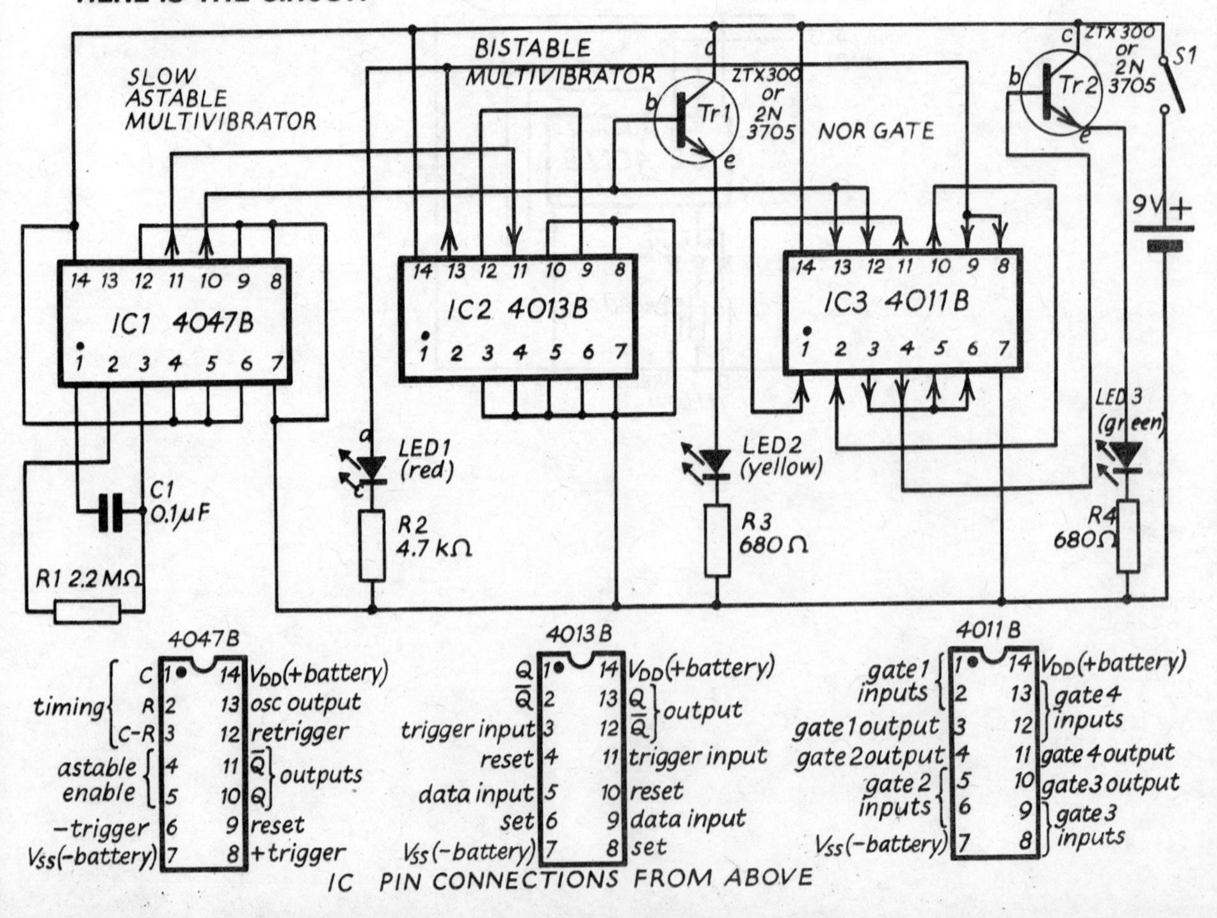

HOW TO BUILD IT

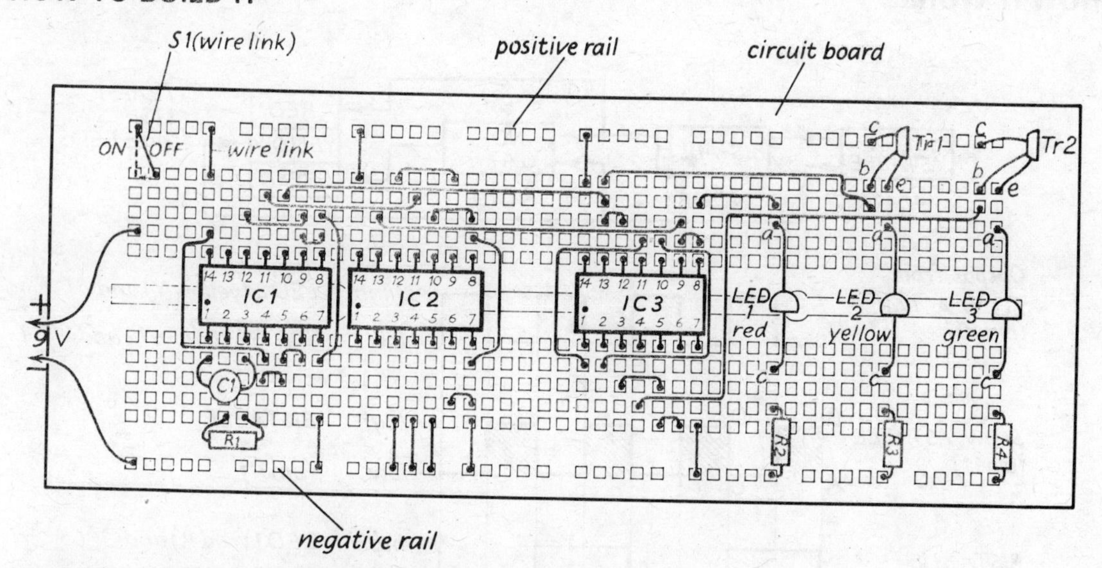

1 Identify pin 1 on the ICs from the small dot or notch at one end of the case. Carefully push each IC into the circuit board in the positions shown, taking care not to bend or touch any of the pins in the process.

2 Insert the wire links from each IC to the positive and negative rails and between other sockets, as shown.

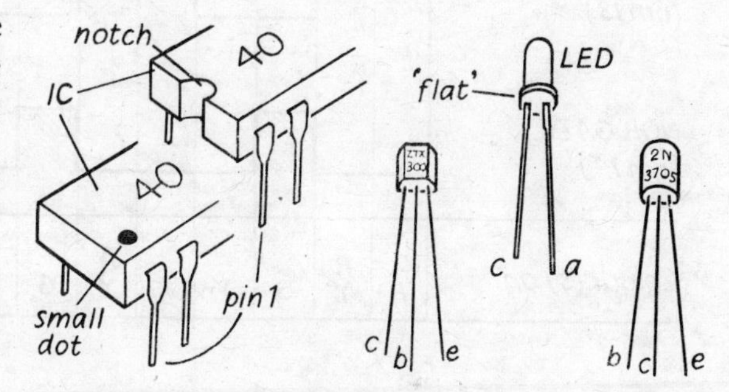

3 Insert $C1$, $R1$, $R2$, $R3$ and $R4$.

4 Insert the three LEDs remembering that the cathode is nearest the 'flat' at the base of the case (or the cathode lead *c* may be shorter than the anode lead *a*).

5 Identify the collector (c), base (b) and emitter (e) leads on the transistors. Insert them in the circuit board, making sure you have them exactly as shown (for ZTX300s) and that the leads are not touching one another.

6 CHECK THE CIRCUIT CAREFULLY.

7 Connect the battery with the correct polarity (the wire link $S1$ acts as the battery on/off switch). The LEDs should light in the order

red	(LED 1)
red and yellow	(LEDs 1 and 2)
green	(LED 3)
yellow	(LED 2)
red	(LED 1) and so on.

If any LEDs don't light when they should, they may be wrongly connected.

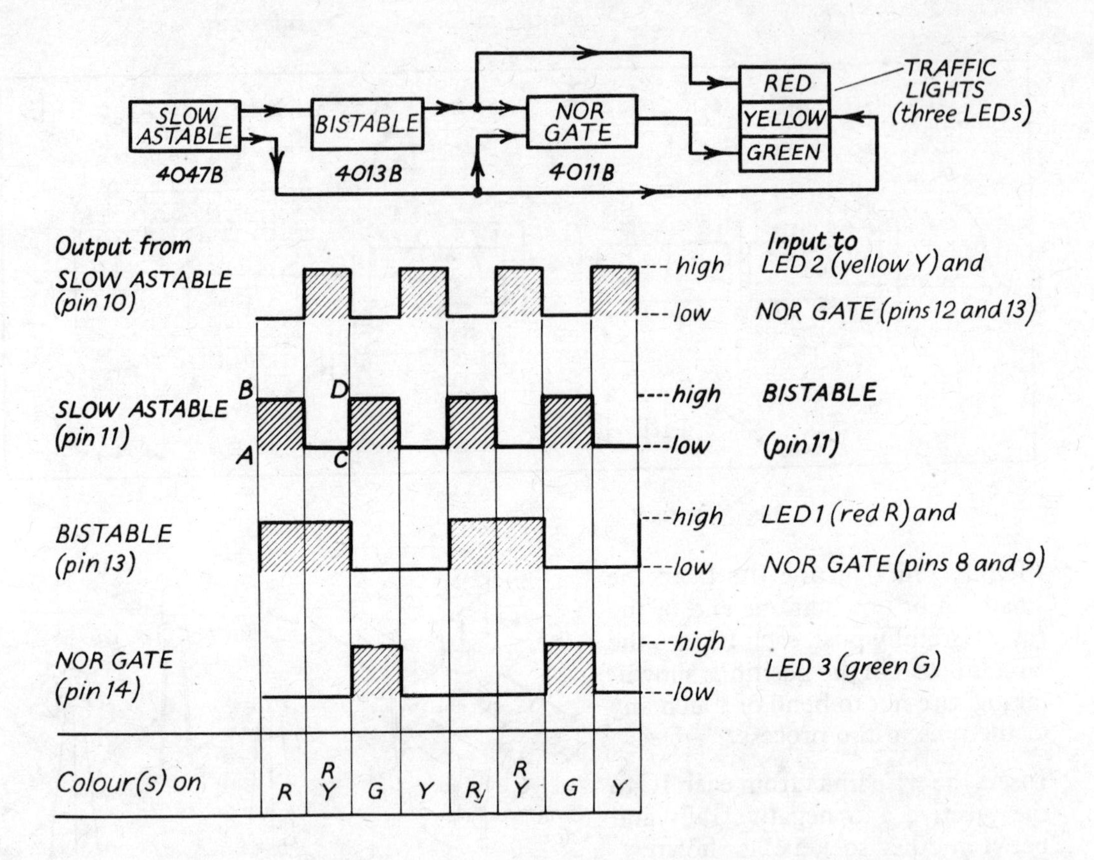

In the diagram above, the first line shows the square pulses from one of the outputs of the astable (the Q output from pin 10). The rate at which these are produced depends on the values of *R*1 and *C*1 (see page 14). The pulses provide the input (via *Tr*1 to prevent overloading) to LED 2 (yellow), making it light up every time they go 'high'. They are also fed to one input of the NOR gate (pins 12 and 13).

The second line shows the square pulses from the other output of the astable (the complementary or $\bar{Q}$ output from pin 11 – see page 14). These go to the input of the bistable (pin 11) and the *positive* edge of each one triggers the bistable, causing its output to switch from 'low' to 'high' or vice versa. For example, AB on the first pulse makes the bistable output go 'high' and remain 'high' until CD, on the second pulse, arrives and switches it from 'high' to 'low'.

The third line shows the output from the bistable (pin 13): the pulses occur at only half the rate of those arriving from the astable but each one is 'high' or 'low' for longer. The output of the bistable is fed to LED 1 (red), making it light up when the output is 'high'. It is also connected to the second input of the NOR gate (pins 8 and 9).

A NOR gate 'opens' and gives a 'high' output only when *both* its inputs are 'low' (see page 19). If you look at the pulses of the first and third lines in the diagram (the coloured ones), you can see why the fourth line gives the output of the NOR gate (at pin 4) – it is 'high' only when the outputs from the bistable and the astable (Q output) are both 'low'. The output from the NOR gate feeds *Tr*2 which drives LED 3 (green) and makes it flash whenever the NOR gate output is 'high'.

The fifth line in the diagram shows the colours that are on at each stage. The 'truth table' (see page 19) for the system is given below

Stage	Red R	Yellow Y	Green G
1	1	0	0
2	1	1	0
3	0	0	1
4	0	1	0
5	1	0	0

(*Note:* the NOR gate is obtained by connecting the four NAND gates in the 4011B – see page 19).

THINGS TO TRY

1 *Action in slow motion.* Change *R*1 from 2.2 MΩ to 10 MΩ. The stages occur more slowly.

2 *Prediction and test.* Predict what will happen if you trigger the bistable from the Q output of the astable rather than the $\bar{Q}$ output. Remember that the bistable triggers on the positive (rising) edge of the input pulse. (*Hint:* you might find the diagram on the opposite page useful.) Check your prediction by altering the circuit.

9 PULSED BLEEPER

This is an extension of the Pulsed Flashing Lamp project, but instead of flashes of light, it produces bleeps of sound, like the time signal 'pips' on radio or the 'bleeps' at a Pelican pedestrian crossing.

WHAT YOU NEED

Dual monostable/astable multivibrator IC (556); NAND gate IC (CMOS 4011B); astable multivibrator IC (CMOS 4047B); resistors – 100 Ω (brown black brown), 1 kΩ (brown black red), 10 kΩ (brown black orange), 1 MΩ (brown black green), 2.2 MΩ (red red green), 10 MΩ (brown black blue); electrolytic capacitor – 10 μF; disc ceramic capacitors – two 0.01 μF, two 0.1 μF; two npn transistors (ZTX300 or 2N3705); loudspeaker 2½ in, 25 to 80 Ω; miniature slide switch SPDT; 9 V battery; battery clip connector; circuit board; PVC-covered tinned copper wire 1/0.6 mm.

HERE IS THE CIRCUIT

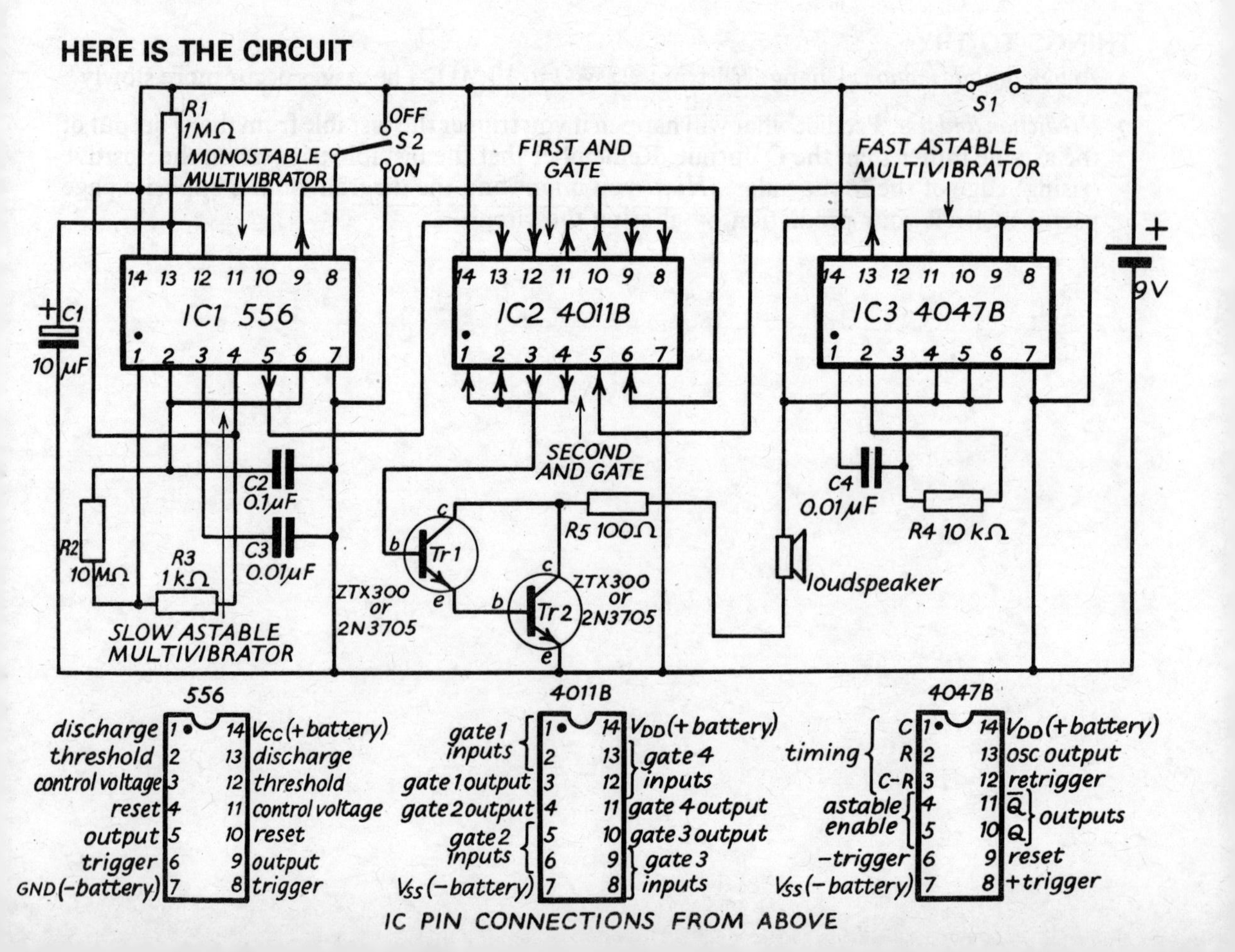

HOW TO BUILD IT

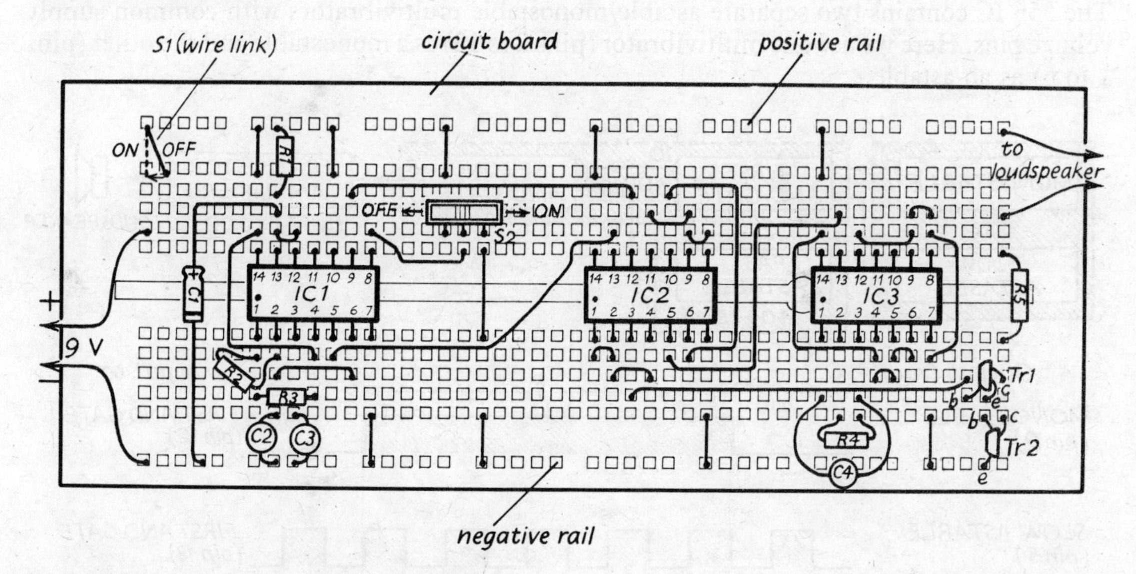

1 Identify pin 1 on the ICs from the small dot or notch at one end of the case. Carefully push each IC into the circuit board in the positions shown, taking care not to bend any of the pins nor to touch those on the CMOS ICs.

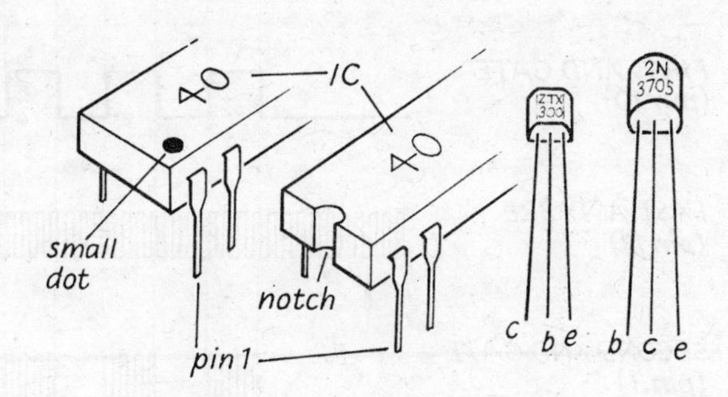

2 Insert wire links from each IC to the positive and negative rails and between other sockets, as shown.

3 Insert *R*1, *R*2, *R*3, *R*4, *R*5, *C*1, *C*2, *C*3, *C*4 and *S*2. Be sure that *C*1 is connected the correct way round, the + end (it has a groove) should go to the bottom of *R*1.

4 Identify the collector (c), base (b) and emitter (e) leads on the transistors. Insert them in the circuit board, making sure you have them exactly as shown (for ZTX300s). Also connect the loudspeaker.

5 CHECK THE CIRCUIT CAREFULLY.

6 Connect the battery with the correct polarity (the wire link *S*1 acts as the battery on/off switch). Switch *S*1 on, then *S*2 on *and* off. (Switch *S*2 off immediately after you have switched it on, otherwise leaving *S*2 on produces continuous bleeping.) You should hear about ten bleeps if all is well. Every time you switch *S*2 on *and* off, the loudspeaker is 'pulsed' to produce a set of bleeps.

HOW IT WORKS

The 556 IC contains two separate astable/monostable multivibrators with common supply voltage pins. Here we use one multivibrator (pins 8 to 13) as a monostable and the other (pins 1 to 6) as an astable.

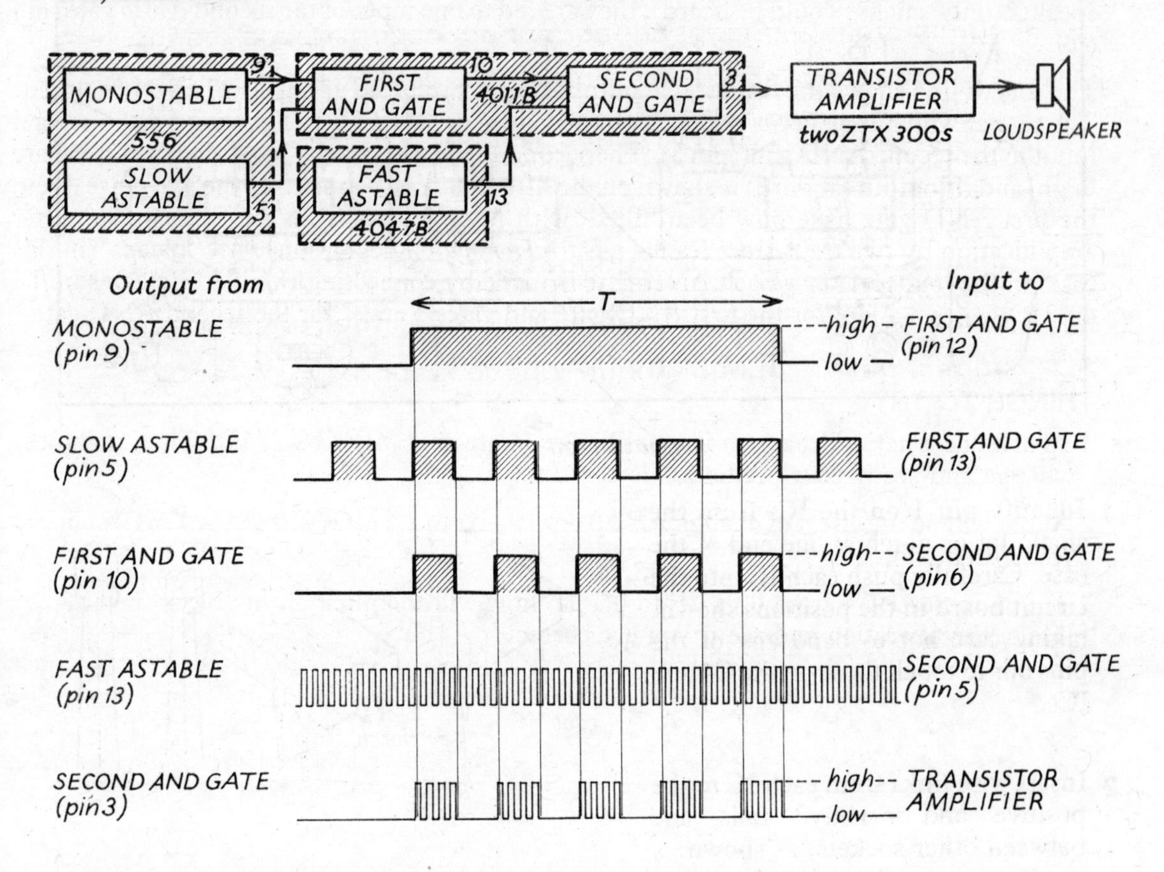

The first line of the diagram above shows the monostable output pulse (at pin 9) which is produced when the monostable is triggered (at pin 8) by the *negative* (falling) edge of a pulse caused by switching $S2$ on and off. With $S1$ on and $S2$ off, pin 8 is at 9 V, but on switching $S2$ on as well, pin 8 becomes connected to 0 V. The negative edge AB so produced provides the triggering for the monostable. The trigger voltage at pin 8 must return to + 9 V before the monostable output pulse ends, otherwise the monostable does not operate, i.e. t (the triggering time) must be less than T (the monostable output pulse time). This means that $S2$ must be switched on and off quickly. The time T of the monostable pulse depends on the values of $R1$ and $C1$ (see page 13). The pulse is fed to one input (pin 12) of the first AND gate.

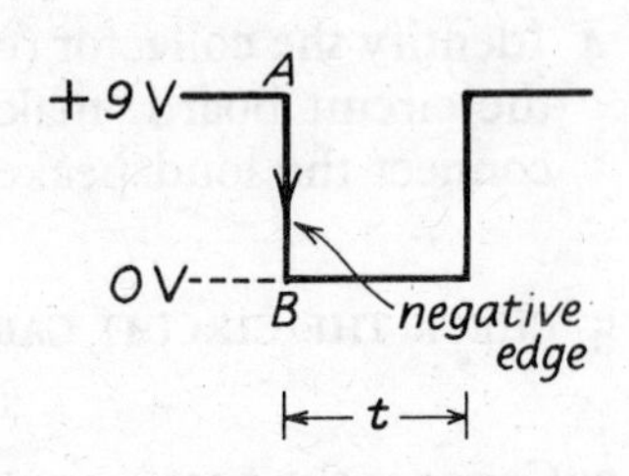

The second line shows the slow astable output pulses (at pin 5) when $S1$ is closed. The rate at which they occur depends on the values of $R2$, $C2$ and $R3$ (see page 11). They supply the second input to the first AND gate (pin 13).

An AND gate 'opens' and gives a 'high' output only when *both* its inputs are 'high' (see page 19). From the first and second lines you can see why the third line gives the output of the first AND gate (at pin 10) – it is 'high' when both inputs are 'high'. In the diagram this occurs five times during the monostable pulse. The pulses are too slow to produce a note in a loudspeaker (only 'clicks' would be heard); they are fed to one input of the second AND gate (pin 6).

The fourth line shows the fast astable output pulses produced at pin 13 when $S1$ is closed. The rate at which they occur depends on the values of $R4$ and $C4$. They supply the second input to the second AND gate (pin 5). The second AND gate 'opens' when *both* its inputs are 'high' and its output (at pin 3) is shown on the fifth line. You can see that the five pulses from the first AND gate have now been 'filled' with faster pulses from the fast astable. After amplification by two transistors (called a *Darlington pair*) they produce five 'bleeps' (in this case) in the loudspeaker. (Each AND gate is made by connecting two NAND gates in the 4011B – gates 3 and 4 for the first AND gate and gates 1 and 2 for the second AND gate.)

THINGS TO TRY

1 Work out what will happen if you change $R1$ from 1 MΩ to 2.2 MΩ. Then make the change and see if your prediction is correct.

2 What will be the effect of changing $R2$ from 10 MΩ to 2.2 MΩ (with $R1 = 1$ MΩ)? Check your prediction.

3 If you increase $C4$ from 0.01 μF to 0.1 μF how will the pitch of the 'bleep' be altered? Make the alteration and see if you are correct.

10 FOUR-BIT BINARY COUNTER

In everyday life counting is done on a scale of ten or decimal system. Computers count on a scale of two or binary system, only using the digits 0 and 1. This circuit counts up to 15 and the total is shown on four LEDs as four binary digits (called *bits*).

WHAT YOU NEED

Astable multivibrator IC (CMOS 4047B); binary counter IC (CMOS 4516B); five LEDs (see *Parts list*, p. 63); resistors – four 680 Ω (blue grey brown), 4.7 kΩ (yellow violet red), 2.2 MΩ (red red green); disc ceramic capacitor – 0.1 μF; 9 V battery; battery clip connector; circuit board; PVC-covered tinned copper wire 1/0.6 mm.

HERE IS THE CIRCUIT

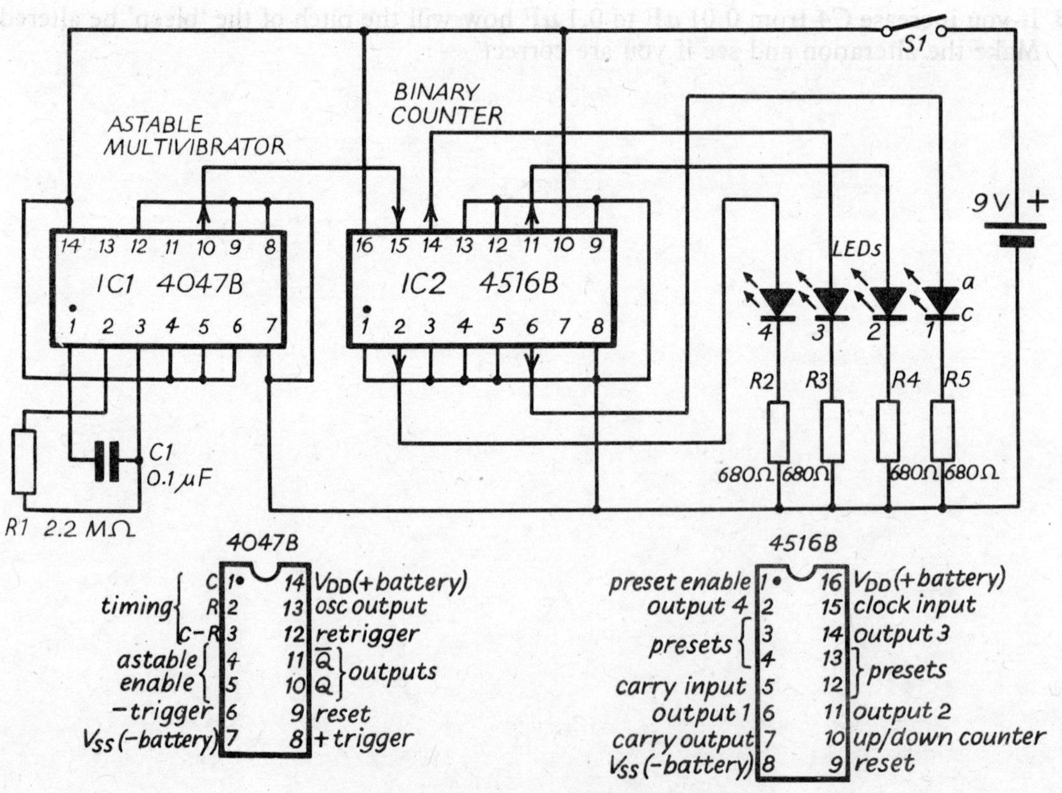

IC PIN CONNECTIONS FROM ABOVE

HOW TO BUILD IT

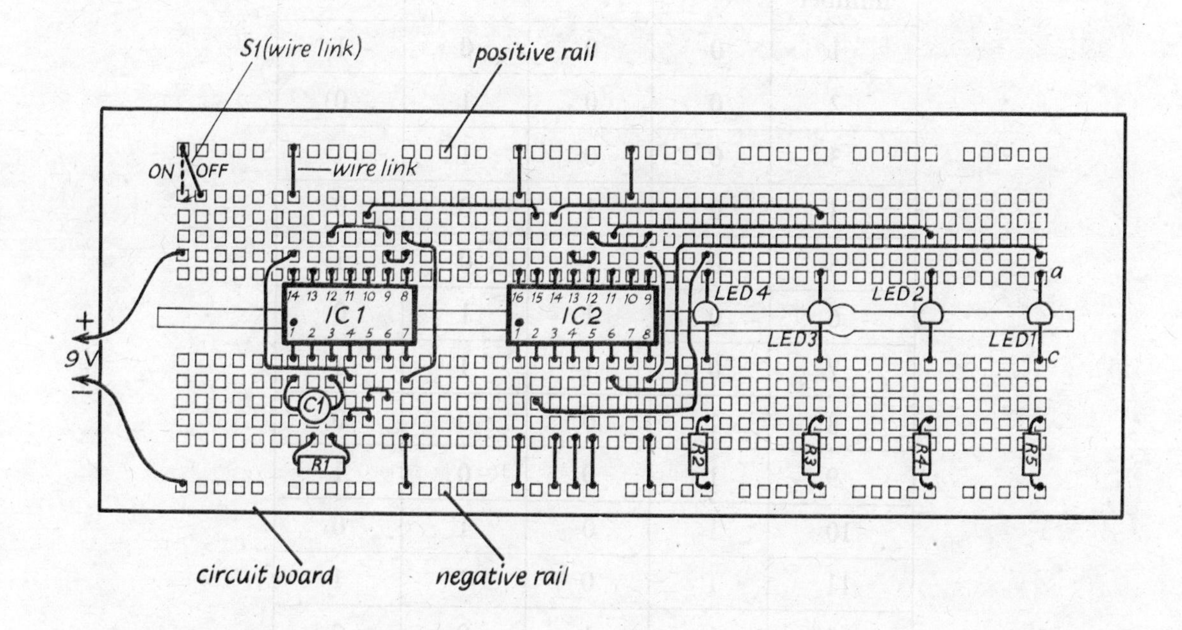

1 Identify pin 1 on the ICs from the small dot or notch at one end of the case. Carefully push each IC into the circuit board in the position shown, taking care not to bend or touch any of the pins.

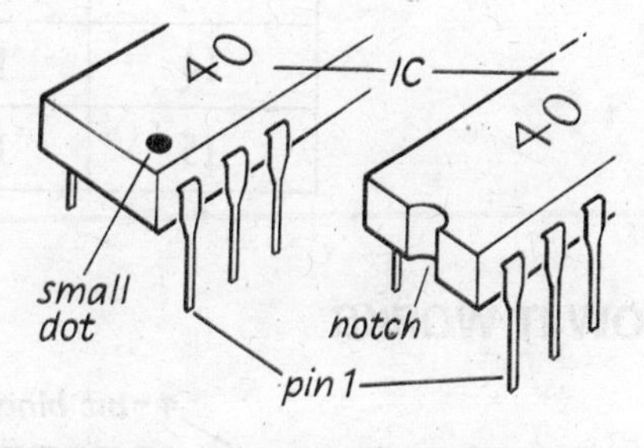

2 Insert wire links from each IC to the positive and negative rails and between other sockets, as shown.

3 Insert *R*1, *R*2, *R*3, *R*4, *R*5 and *C*1.

4 Insert LEDs 1 to 4 remembering that the cathode is next to the 'flat' at the bottom of the plastic case (or the cathode lead *c* may be shorter than the anode lead *a*).

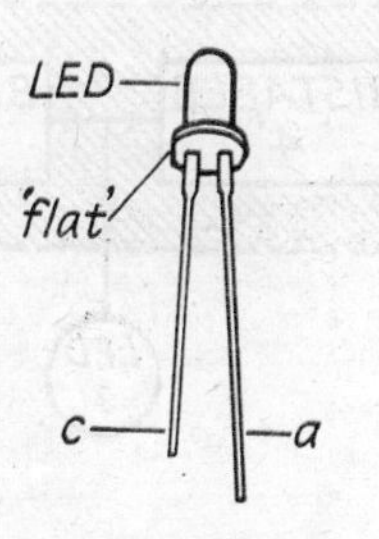

5 CHECK THE CIRCUIT CAREFULLY.

6 Connect the battery with the correct polarity (the small wire link acts as the battery on/off switch *S*1). Switch *S*1 on. The astable feeds 'slow' pulses into the binary counter and the LEDs should light up in the order shown in the table below – a lit LED is represented by figure 1 and an unlit one by figure 0.

The first pulse (1 in binary) lights LED 1, the second pulse (10 in binary) lights LED 2 and puts LED 1 out, the third pulse (11 in binary) keeps LED 2 alight and lights LED 1 again, the fourth pulse (100 in binary) lights LED 3 but puts out LEDs 1 and 2 and so on until at the fifteenth pulse (1111 in binary) all four LEDs are alight.

Pulse number	LED 4	LED 3	LED 2	LED 1
1	0	0	0	1
2	0	0	1	0
3	0	0	1	1
4	0	1	0	0
5	0	1	0	1
6	0	1	1	0
7	0	1	1	1
8	1	0	0	0
9	1	0	0	1
10	1	0	1	0
11	1	0	1	1
12	1	1	0	0
13	1	1	0	1
14	1	1	1	0
15	1	1	1	1

HOW IT WORKS

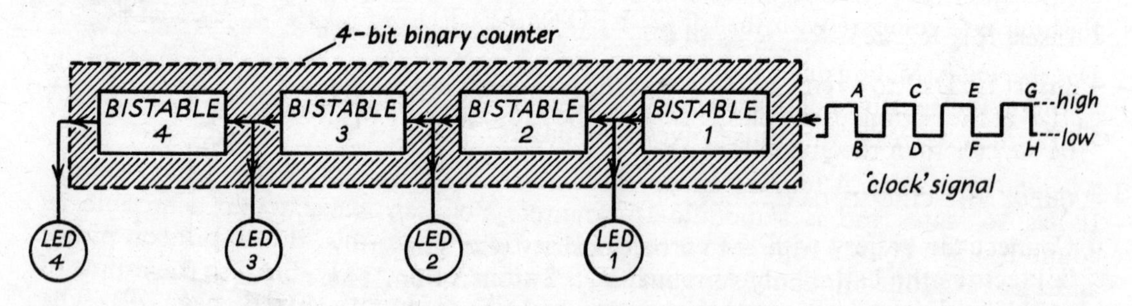

A four-bit binary counter contains four bistables. It is driven by 'slow' square wave pulses from an astable, called the 'clock'. In the diagram above, the output of bistable 1 (BS1) drives BS2 which then drives BS3, which in turn drives BS4. In the 4516B the bistables are not connected quite so simply but they give the same end-result and for the purpose of explaining the action of a binary counter it is easier to consider the arrangement shown.

If the outputs of all bistables are initially 'low', the output condition of the counter in binary form is 0000. When the first clock pulse arrives, BS1 changes state from 'low' to 'high' on the *negative-going* part AB of the pulse. At the end of the first pulse the output condition is 0001, i.e. LED 1 only is alight. The output of BS1 has gone from 'low' to 'high' – a positive change. When this is fed to BS2, it has no effect because it is not a negative-going change and so LED 2 (as well as 3 and 4) stay off.

The *negative-going* part CD of the second clock pulse changes the output of BS1 again but this time from 'high' to 'low', i.e. a negative change, and LED 1 goes off. This negative-going change triggers BS2 whose output goes from 'low' to 'high', i.e. LED 2 comes on. The positive-going change in the output of BS2 does not affect BS3 and at the end of the second pulse the output condition of the counter is 0010.

The *negative edge* EF of the third clock pulse changes the output of BS1 from 'low' to 'high' and this positive-going change leaves the other bistables unchanged and gives an output condition of 0011, i.e. LEDs 1 and 2 alight.

The edge GH of the fourth clock pulse changes BS1 from 'high' to 'low', which in turn (being a negative edge) causes BS3 to go from 'low' to 'high'. The output condition of the counter is then 0100, i.e. LED 3 only alight.

The output condition of the four bistables for the first four clock pulses is 0001, 0010, 0011, 0100 (i.e. the binary numbers for 1, 2, 3, 4). Comparison with the table on page 54 will show that it always agrees with the binary numbers for the remaining pulses up to 15.

Note. The 4516B is a synchronous-connected counter and its action is different from that of the simpler asynchronous-connected arrangement shown opposite; for example it switches on the positive-going part of clock input pulses.

THINGS TO TRY

1 *Clock signal.* Connect another LED in series with a suitable current-limiting resistor (4.7 kΩ) between the output of the astable (pin 10) and the negative rail. How does its flashing rate compare with that of (*a*) LED 1, (*b*) LED 2, (*c*) LED 3, (*d*) LED 4?

2 *Down-counter.* Make a down-counter instead of an up-counter by connecting pin 10 on the 4516B to the negative rail instead of to the positive rail. The binary numbers shown by the LEDs then decrease with successive clock pulses.

3 *Modulo-10 counter.* A four-bit binary counter counts from 0 to 15 then resets to zero. It has 16 states and is a modulo-16 counter. You can make it into a modulo-10 counter which counts from 0 to 9 before resetting. To do this, the outputs on pins 2 and 11 of IC2 must not only supply LEDS 4 and 2 respectively, but also the inputs of a two-input AND gate (made from two NAND gates in a 4011B, page 19). The output of the AND gate is connected to pin 9 (reset) on IC2 which must be disconnected from the negative rail in this case.

Try it out but drive LEDS 4 and 2 with two transistors as is done in project 8 (page 44) by Tr 1 for the yellow LED.

Can you explain why the count stops at 9?

11 REACTION TIMER

How quickly do you respond to a signal? With this gadget you can get a measure of the time you take to react to a light coming on. With practice you might even improve. You could also compare your performance with that of other people.

WHAT YOU NEED

Astable multivibrator IC (CMOS 4047B); BCD counter IC (CMOS 4510B); BCD decoder driver IC (CMOS 4511B); 7-segment LED display, common cathode (see *Parts list*, p. 63); resistors – 220 Ω (red red brown), two 100 kΩ (brown black yellow), 220 kΩ (red red yellow), 1 MΩ (brown black green), 2.2 MΩ (red red green); disc ceramic capacitor – 0.1 μF; miniature slide switch SPDT; 9 V battery; battery clip connector; circuit board; PVC-covered tinned copper wire 1/0.6 mm.

HERE IS THE CIRCUIT

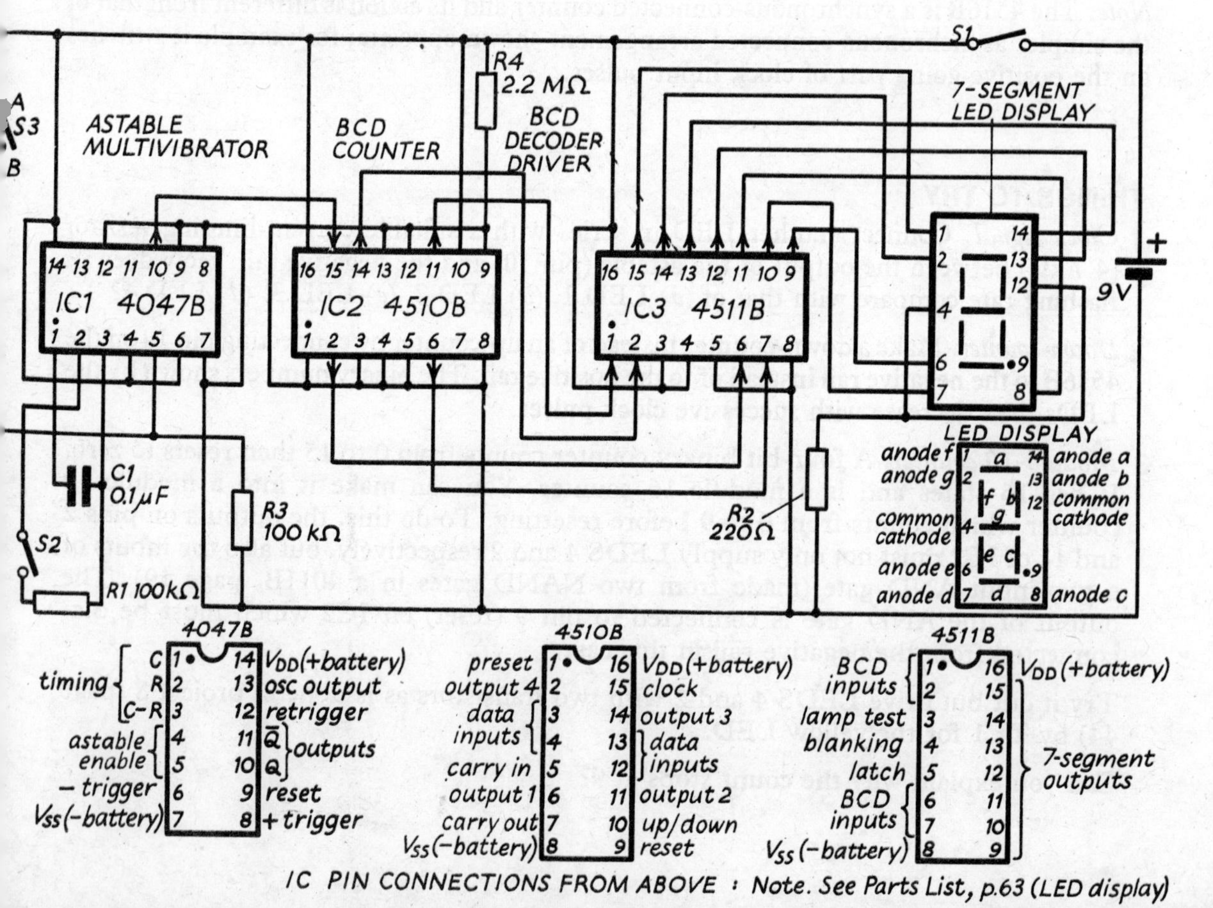

HOW TO BUILD IT

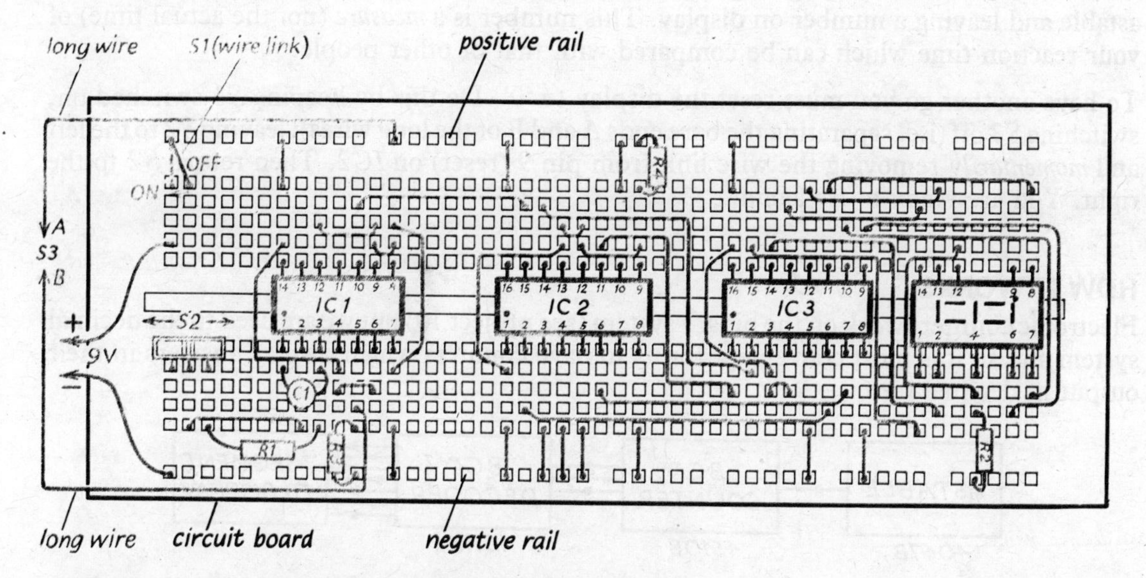

1 Identify pin 1 on the ICs from the small dot or notch at one end of the case. Carefully push each IC into the circuit board in the position shown, taking care not to bend or touch any of the pins.

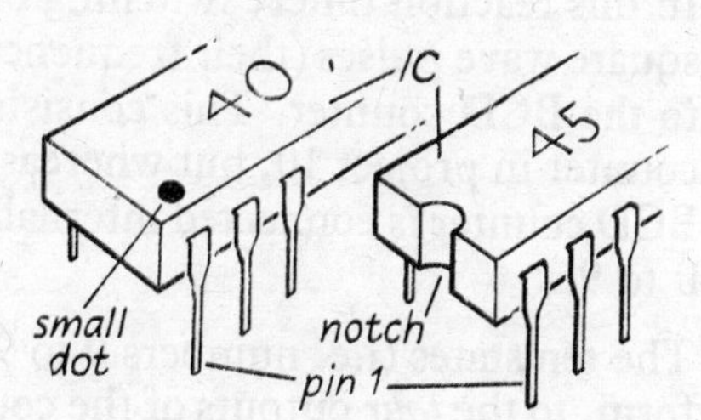

2 Carefully insert the 7-segment LED display in the board as shown – note that four of its pins are missing.

3 Insert wire links from each IC to the positive and negative rails and between other sockets, as shown. Make the wires from A and B at least 30 cm long.

4 Insert *R*1, *R*2, *C*1 and *S*2.

5 CHECK THE CIRCUIT CAREFULLY.

6 Connect the battery with the correct polarity *but do not switch wire link S*1 (the battery on/off switch) *on yet.*

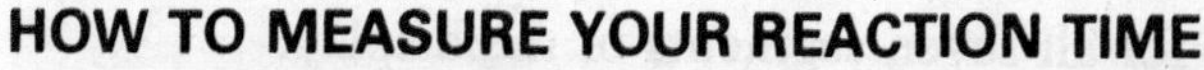

HOW TO MEASURE YOUR REACTION TIME

First you have to make sure that the 7-segment display starts from '0'. To do this have *S*2 to the left, switch *S*1 on, keep the bare ends A and B of the two long wires apart (i.e. *S*3 is off) and remove the wire link connecting pins 9 (reset) and 12 on *IC*2. When '0' appears on the display return the link. You are now ready to start.

Push *S*2 to the right. Get someone to hold wires A and B and, *unseen by you*, to bring and keep the two bare ends together (e.g. by doing it below the table where you may both be sitting). This switches *S*3 on and sets the astable working – its pulses being counted by the LED display.

In the meantime you have to watch the display and at the same time have your finger on $S2$. *As soon as you see the display lighting up, switch S2 to the left as fast as you can* so turning off the astable and leaving a number on display. This number is a *measure* (not the actual time) of your reaction time which can be compared with that of other people.

To have another go you must reset the display to '0'. Do this by keeping $S1$ switched on, switching $S3$ off (i.e. separating the bare ends A and B of the long wires), leaving $S2$ to the left and *momentarily* removing the wire link from pin 9 (reset) on $IC2$. Then return $S2$ to the right. You are ready to try again.

HOW IT WORKS

Electronic counters work on the binary system (see project 10) but we are used to the decimal system and so it is most convenient for us to have counters which display answers in their output in decimal form.

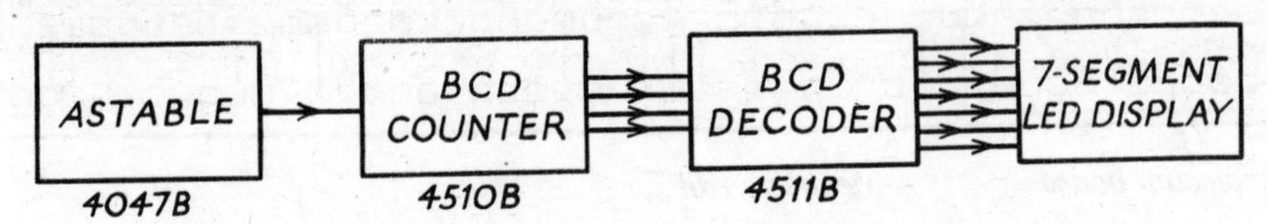

In this reaction timer, switching on $S3$ 'enables' the astable (see page 15) to send fairly fast square wave pulses (their frequency is determined by the values of $R1$ and $C1$ – see page 14) to the BCD counter. This consists of four bistables (among other things), like the binary counter in project 10, but whereas it counts from 0 to 15, i.e. has sixteen output states, the BCD counter is connected internally so that it only uses *ten* of these states, i.e. it counts from 0 to 9.

The ten states (i.e. numbers 0 to 9) are fed in turn (as each astable pulse arrives), in binary form, to the *four* outputs of the counter. For example, when the count is 3 (in binary 0011), outputs one and two (pins 6 and 11) are 'high' and outputs three and four (pins 14 and 2) are 'low'. Or again, when the count is 9 (in binary 1001), outputs one and four are 'high' and outputs two and three are 'low' and so on. (This is why it is called a BCD or Binary Coded Decimal counter – it represents decimal numbers in binary form.)

The BCD decoder receives the four outputs from the counter at its *four* inputs (pins 1, 2, 6 and 7) and has to convert them into a form which makes a decimal display possible. The 7-segment LED display consists of seven small LEDs which produce the numbers 0 to 9 when various combinations of the seven segments light up. The BCD decoder therefore has to create *seven* outputs (pins 9 to 15) from its four inputs. This it does and feeds to the display.

$R2$ is a common current-limiting resistor for all seven segments of the display.

THINGS TO TRY

1 *'Reducing' your reaction time*. Change $R1$ from 100 kΩ to 220 kΩ. You should now find that your reaction time as shown on the display is less! This is due to the astable sending pulses more slowly to the counter.

Replace the 220 kΩ by 1 MΩ and you will see the counter working in 'slow motion' when $S1$ and $S3$ are on and $S2$ is to the right.

2 *Using a binary counter instead of a BCD counter.* Try to work out and then check what happens when you replace the 4510B BCD counter in the reaction timer by the 4516B Binary counter (which counts from 0 to 15) from project 10.

3 *For the enthusiastic.* Rebuild the Binary counter of project 10 but replace the 4516B by the 4510B BCD counter. What would you expect to be the result? Would it count up to 15 as before? If not, to what?

4 *BCD counter to 99.* You need a second 4510B, connected to another 4511B which supplies a 'tens' 7-segment LED display. The 'carry out' of the first 4510B (pin 7) is connected to the 'clock' of the second 4510B (pin 15). Otherwise all connections are as before.

12 MW/LW RADIO RECEIVER

If you enjoy making things that work you will get a lot of pleasure from building this little 'portable' radio. It picks up both medium and long wave stations on a ferrite rod aerial and gives loudspeaker reception.

WHAT YOU NEED

TRF radio IC (ZN414Z); audio amplifier IC (LM380); two npn transistors (ZTX300 or 2N3705); resistors – 330 Ω (orange orange brown), 1.5 kΩ (brown green red), three 10 kΩ (brown black orange), 100 kΩ (brown black yellow), 470 kΩ (yellow violet yellow); variable capacitor 0.0005 μF; electrolytic capacitors – two 1 μF, 100 μF; disc ceramic capacitors – two 0.01 μF, two 0.1 μF; loudspeaker 2½ in, 25 to 80 Ω; ferrite rod 100 mm × 9 mm; knob; 9 V battery; battery clip connector; circuit board; 10 metres enamelled copper wire 30 gauge; PVC-covered tinned copper wire 1/0.6 mm; 2 mm bore plastic sleeving; four small rubber bands.

HERE IS THE CIRCUIT

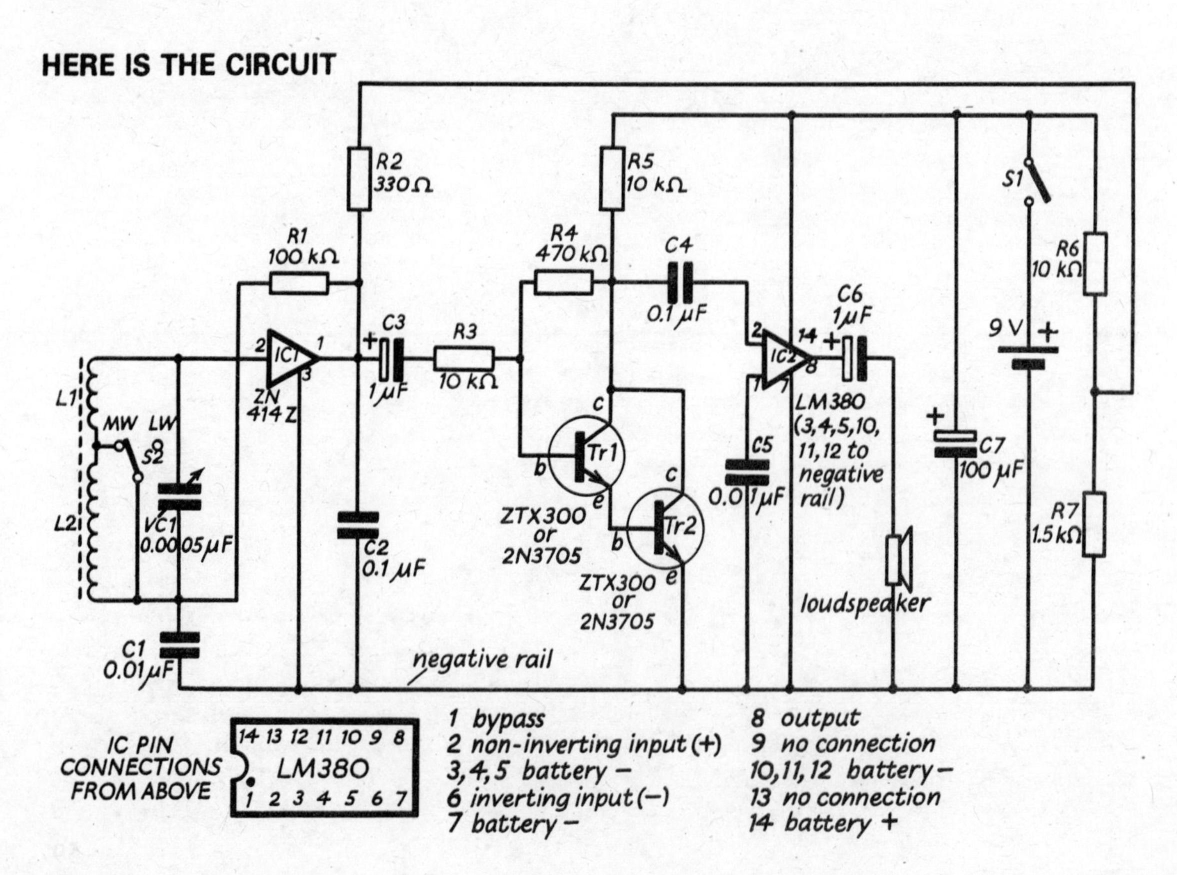

HOW TO BUILD IT

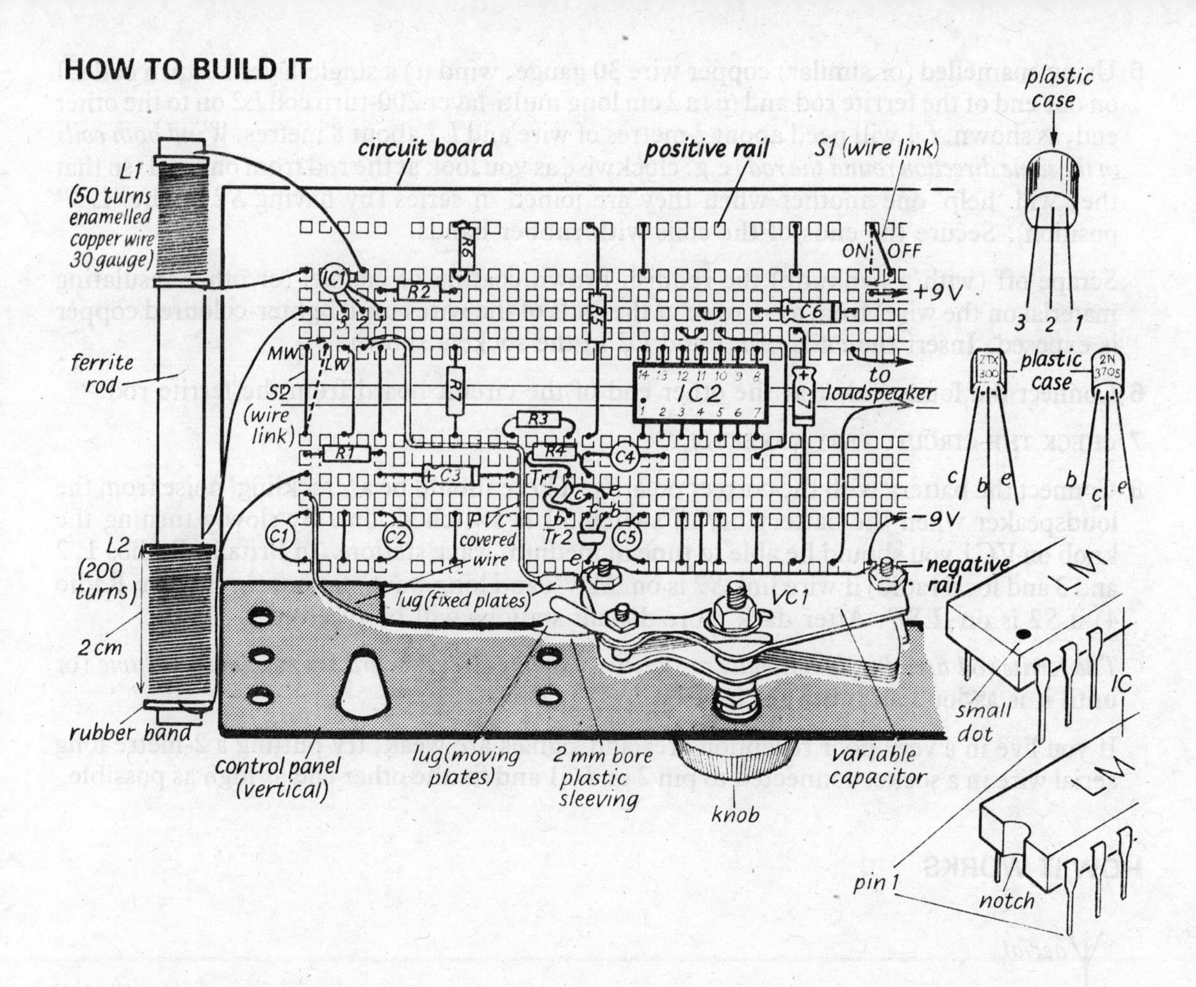

1 Identify (*i*) pins 1, 2 and 3 on the ZN414Z (*ii*) the collector (c), base (b) and the emitter (e) leads on the transistors (ZTX300 or 2N3705) and (*iii*) pin 1 on the LM380. Carefully push each device into the circuit board in the positions shown, making sure that none of the leads on the *IC*1, *Tr*1 and *Tr*2 are touching where they come out of the bottom of the case.

2 Insert wire links from the ICs to the positive and negative rails and between other sockets, as shown.

3 Insert *R*1 to *R*7 and *C*1 to *C*7. Be sure that the electrolytic capacitors *C*3, *C*6 and *C*7 are connected the correct way round, the + end has a groove and the − end a black band round it (usually).

4 Cut two 8 cm lengths of PVC-covered tinned copper wire and remove the insulation from about 1 cm of each end. 'Join' one end of each wire to a lug on the variable capacitor *VC*1, using a short piece of 2 mm bore plastic sleeving. (See page 8 for making 'joints'.) Fix *VC*1 securely to the control panel by tightening its locking nut with pliers and put the knob on the spindle. Slot the panel into the circuit board. Connect the wire from the fixed plates lug to a socket joined to pin 2 on *IC*1 and the wire from the moving plates lug to a socket joined to both *R*1 and *C*1.

5 Using enamelled (or similar) copper wire 30 gauge, wind (*i*) a single-layer 50-turn coil $L1$ on one end of the ferrite rod and (*ii*) a 2 cm long multi-layer 200-turn coil $L2$ on to the other end, as shown. $L1$ will need about 2 metres of wire and $L2$ about 8 metres. *Wind both coils in the same direction round the rod* (e.g. clockwise as you look at the rod from one end) so that they will 'help' one another when they are joined in series (by having $S2$ in the 'LW' position). Secure the ends of the coils with rubber bands.

Scrape off (with a penknife) the reddish brown coating of enamel (or other insulating material on the wire) from a cm or so of the ends of the wires until lighter-coloured copper is exposed. Insert the ends of $L1$ and $L2$ in the sockets shown.

6 Connect the loudspeaker at the other end of the circuit board from the ferrite rod.

7 CHECK THE CIRCUIT VERY CAREFULLY.

8 Connect the battery with the correct polarity. There should be a 'crackling' noise from the loudspeaker when the battery on/off switch $S1$ is switched 'on'. By slowly turning the knob on $VC1$ you should be able to tune in medium wave stations (in Britain, Radios 1, 2 and 3 and local radio) if wire link $S2$ is on 'MW', and long wave stations (in Britain, Radio 4) if $S2$ is on 'LW'. After dark more distant stations will be received.

The ferrite rod aerial is directional, so slowly rotate the circuit board for maximum volume (or until it is as loud as you want it to be).

If you live in a very poor reception area and signals are weak, try putting a 2-metre long aerial wire in a socket connected to pin 2 on $IC1$ and fix the other end as high as possible.

HOW IT WORKS

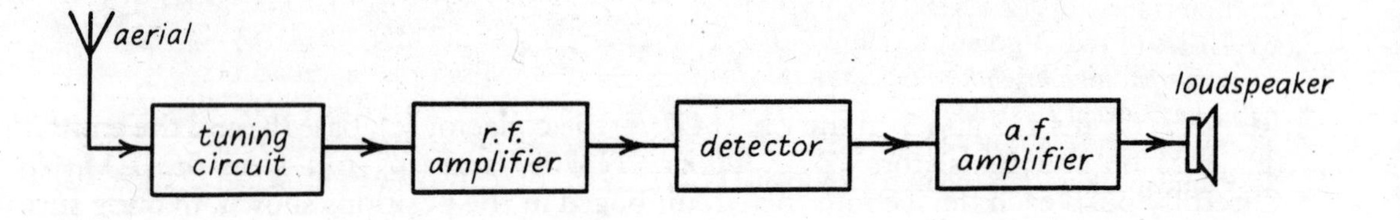

The ferrite rod acts as the aerial. The tuning circuit is $VC1$ with $L1$ on medium waves and with $L1$ and $L2$ in series on long waves. $IC1$ is both r.f. (radio frequency) amplifier and detector. The a.f (audio frequency) amplifier consists of $Tr1$ and $Tr2$ as a preamplifier and $IC2$ as a power amplifier driving the loudspeaker. (If you want to know more about 'How Radio Works' you will find an explanation in the companion book *Adventures with Electronics*.)

The ZN414Z ($IC1$) works off 1.2 to 1.6 V: it is obtained here by having $R6$ and $R7$ connected as a voltage divider across the battery (9V) tapping off from $R7$.

PARTS LIST

- 1 circuit board (e.g. *Experimentor 300, Bimboard, Balinu* or *Prototype* board, *ACE*, *Superstrip*, etc.)†
- 1 dual monostable/astable IC (556)
- 1 CMOS monostable/astable IC (4047B or 14047B)
- 1 CMOS D-type bistable IC (4013B or 14013B)
- 1 CMOS quad 2-input NAND gate IC (4011B or 14011B)
- 1 CMOS decade counter IC (4017B or 14017B)
- 1 CMOS binary counter IC (4516B or 14516B)
- 1 CMOS BCD counter IC (4510B or 14510B)
- 1 CMOS BCD decoder-driver IC (4511B or 14511B)
- 1 7-segment LED display, common cathode – *all do not have the pin connections shown on p. 56.*
- 1 audio amplifier IC (LM380)
- 1 *TRF (tuned radio frequency) radio IC (ZN414Z)
- 2 npn transistors (ZTX300 or 2N3705)
- 6 LEDs (4 red, 1 green, 1 yellow) – *low current high brightness types suitable for low current ICs*
- 1 *photocell (e.g. ORP12)
- 1 *loudspeaker 2½ in (63 mm), 25 to 80 Ω
- 1 *variable capacitor 0.0005 μF (Jackson Dilecon)
- 1 *ferrite rod 100 mm × 9 mm
- 1 *knob
- 1 miniature p.c.b. (printed circuit board) slide switch SPDT (single pole double throw)
- 1 push-button on/off switch (optional)
- 5 m PVC-covered tinned copper wire 1/0.6 mm
- 10 m enamelled copper wire 30 gauge (diameter 0.3 mm)
- 10 cm *plastic sleeving 2 mm bore
- 50 cm *plastic sleeving 1 mm bore (optional)
- 4 small rubber bands
- 1 battery 9 V (e.g. PP3)
- 1 battery clip connector
- 42 resistors, carbon, ½ watt
  *100 Ω, 220 Ω, *330 Ω, six 680 Ω, *two 1 kΩ, 1.5 kΩ, *3.9 kΩ, *4.7 kΩ, two 8.2 kΩ *three 10 kΩ, 12 kΩ, four 15 kΩ, 18 kΩ, two 22 kΩ, two 27 kΩ, *33 kΩ, two 39 kΩ, two 47 kΩ, *two 100 kΩ, 220 kΩ, *470 kΩ, 1 MΩ, two 2.2 MΩ, 10 MΩ.
- 4 disc ceramic capacitors
  *two 0.01 μF, *two 0.1 μF
- 5 electrolytic capacitors
  two 1 μF, 4.7 μF, *10 μF, *100 μF

*Items used in *Adventures with Electronics*.
†Not all of these have a control mounting panel but it is not difficult to improvise one.

FIRST INDIAN EDITION–1993

Distributors:

BPB BOOK CENTRE
376, Old Lajpat Rai Market, Delhi–110006

BUSINESS PROMOTION BUREAU
8/1, Ritchie Street, Mount Road, Madras–600002

BUSINESS PROMOTION BUREAU
4-3-268-C, Giriraj Lane, Bank Street, Hyderabad–500195

COMPUTER BOOK CENTRE
12, Shrungar Complex, M. G. Road, Bangalore–560001

COMPUTER BOOK CENTRE
Kothi No.–535, Sector–7, Panchkula–134109, CHANDIGARH

ADDRESSES

ENGLAND

A complete *Adventures with Microelectronics* kit of good quality parts can be bought from: Unilab Ltd, The Science Park, Hutton Street, Blackburn, England. BB1 3BT (Tel: 0254 681222)

INDIA

COMPONENTS/PCB may be available with VISHA ELECTRONICS 17, Kalpana Building, 349 Lamington Road, Opp. Police Station, Bombay 400 007 (Tel: 3862622, 3862650)

Printed in India under arrangement with
JOHN MURRAY (PUBLISHERS) LTD., ENGLAND

Published by Manish Jain for BPB Publications, B–14, Connaught Place, New Delhi and Printed by him at Pressworks, Delhi.